INNOVATION AND APPLICATION OF GREEN BUILDING
MATERIALS
— RESIDENTIAL BUILDINGS

绿色建筑材料的创新与应用 · 居住建筑
VOL . 2

深圳市艺力文化发展有限公司 编

华南理工大学出版社
SOUTH CHINA UNIVERSITY OF TECHNOLOGY PRESS
· 广州 ·

图书在版编目（CIP）数据

绿色建筑材料的创新与应用 = Innovation and application of green
building materials：汉英对照 / 深圳市艺力文化发展有限公司编 .-- 广州：
华南理工大学出版社，2014.4
ISBN 978-7-5623-4187-1

Ⅰ．①绿… Ⅱ．①深… Ⅲ．①建筑材料 - 无污染技术 - 汉、英 Ⅳ．① TU5

中国版本图书馆 CIP 数据核字（2014）第 044423 号

绿色建筑材料的创新与应用
Innovation and Application of Green Building Materials
深圳市艺力文化发展有限公司 编

出 版 人：韩中伟

出版发行：华南理工大学出版社

（广州五山华南理工大学 17 号楼，邮编 510640）

http://www.scutpress.com.cn　E-mail: scutc13@scut.edu.cn

营销部电话：020-87113487　87111048（传真）

策划编辑：赖淑华

责任编辑：史　册　赖淑华

印 刷 者：深圳市汇亿丰印刷包装有限公司

开　　本：965mm×1440mm　1/8　印张：60

成品尺寸：245mm × 330mm

版　　次：2014 年 4 月第 1 版　2014 年 4 月第 1 次印刷

定　　价：598.00 元（上、下册）

Preface

材料打造独特建筑

在建筑设计中，材料担任了重要的角色。

没有选对正确的材料，即便是优质的设计概念和建筑结构也会受到冲击。而另一方面，如果缺乏优质的建筑结构和空间组织概念，那即便是最好最贵的材料也无法改善建筑本身。

因此，让二者同步于设计之中，既实现建筑功能及其环境空间组织和结构的优质设计，又保证选材的质量，这点在建筑设计中非常重要。

图片 1：离别堂，2009 (Photo @ Tomaz Gregoric)

在建筑的实际施工过程中，建筑往往要根据特定的条件来建造空间和选用材料，这就是建筑设计中的妥协。主要的妥协因素当然是来自于客户所给的预算，选材经常是根据城市规划的需求来确定材料的颜色与类型，通常的限制因素有防火性能、声音参数、传热条件以及在某些异常天气中的特殊材料使用性能，这些限制因素在建筑设计中起了重要作用。一座好的建筑会遵循这些妥协因素，而最后的建筑效果也往往会比那些没有任何限制因素的建筑要好。

在实际施工的过程中，我们倾向于设计纯粹的建筑 —— 空间的创造产生于选址环境与功能性的规划。选择这二者同步的建筑结构，对我们来说影响重大。我们相信，如果建筑外形根据特定的逻辑方法建造，那建筑结构的所有成分和材料选择都能从建造过程的本身中衍生出来。

图片 2：贝特体育馆，2013

当我们建造一栋建筑时，我们会对选址、当地的气候条件、附近建筑的用材及建筑元素，尤其是对过去建筑的用材做调查。我们相信，遵循传统可以创造出最好的可持续效果。使用当地的材料和其历史中特定场所的建材，不仅能让新建筑与周围的环境保持和谐，还能创造出独一无二的建筑。

图片 3：6x11 阿尔卑斯别墅，2009 (Photo @ Tomaz Gregoric)

斯佩拉·维德尼克、罗克·奥曼(OFIS 建筑公司)

Materials That Create Unique Architecture

The materials play a significant role in architecture.

Incorrect choice of materials affects the impact of the building, no matter of the high quality of the concept and volume. But on the other hand, if the volume and the concept of the spatial organisation have no quality, even the best or most expensive materials cannot improve it.

Therefore, the synchronisation of both, which means, achieving spatial organisation, volume with relation to functionality and its surrounding as well as the choice of material, is important.

Picture 1: Farewell chapel, 2009 (Photo @ Tomaz Gregoric)

Picture 1

In architectural practice, the way of generating architecture depends on specific circumstances. So creating the space and choosing materials is a set of compromises. The main compromise is, of course, the budget of the client. Very often, choice of materials depends on urbanistic requirements that define colour and type of materials. Common restrictions, such as fire-resistance, acoustic parameters, thermal conditions or behaviours of specific materials in certain climate, play a significant role. Good architecture follows these compromises, and the final result is better than it would be without restrictions.

In process of our practice, we tend to create buildings that are pure — creation of space arising from the site context and functional plan organisation. It is one of most important impacts for us that the structure of the building is chosen in synchronisation with this two. We believe that if form is generated with certain logic, all components of the building's structure, as well as choice of materials, arise from the process itself.

Picture 2: Bate stadium, 2013

Picture 2

When creating the building, we make a research of the site, such as its climatic conditions as well as what materials and architectural elements are used in close surrounding, especially the ones that were used historically. We believe that following tradition creates the best sustainable effect. Use of local materials and the ones that were used on specific site through the history, not only creates harmonisation of the new build with close environment and context, but also generates architecture that is unique and special.

Picture 3: 6x11 Alpine Hut, 2009 (Photo @ Tomaz Gregoric)

Spela Videcnik and Rok Oman, OFIS arhitekti

Picture 3

Contents

CO2-Saver House

Ilma Grove House

Villa 921

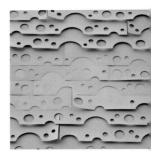

Fabric Facade Almere

CO2—Saver 房子

CO2—Saver House

This simple sustainable house – like a chameleon – blends with its surrounding area on Laka Lake in Upper Silesia in Poland. Colourful planks within the timber facade reflect the tones of the landscape. The window reveals with fibre cement cladding frame images of the countryside. Analogical to the most creatures, the building is outside symmetrical, although the internal zones – according to function – are arranged asymmetrically.

The form of the house is designed to optimise the absorbance of solar energy. Approximately 4/5 of the building envelope is facing to the sun. The single storey living space on the ground floor is externally clad with untreated larch boarding. Solar energy is gained there by the amply glazed patio and stored in the loam walls. Solar thermal collectors are located on the roof and a photovoltaic system is planed for the future. The dark facade of the "black box" – a three storey structure clad with charcoal coloured fibre cement panels – is warmed by the sun, reducing heat loss to the environment. The passive and active solar energy concepts and a high standard of thermal insulation are enhanced by a ventilation plant with thermal recovery system.

The design of the project was determined by the twin goals of low lifecycle costs and a reduction in construction costs. All details are simple, but well thought out. The house did not cost more than a conventional one in Poland.

　　这个简洁的可持续性住宅，像变色龙一样，与波兰上西里西亚的拉卡湖周围环境融为一体。木房正面的各色木材反映了景观的色调。带有纤维水泥外墙的窗户展示了乡村的框型影像。如同大多数生物一样，此房子外部是对称的，但内部空间根据功能非对称设置。

　　房子被设计成充分吸收太阳能的形式，大约 4 / 5 的房子表面是面向太阳的。首层生活空间外面覆盖着未经处理的落叶松木板，太阳能通过玻璃般光滑的露台获得并存储在壤土墙里。太阳能集热器安装在屋顶，未来计划安装光伏系统。"黑匣子"一个三层结构，覆盖着炭黑色纤维水泥板的深色外观由太阳加热，可减少热量流失到环境中。热回收系统的通风装置增强了主导和诱导式太阳能观念和热绝缘材料的标准。

　　该项目的设计有两个目标：低寿命周期成本和低建设成本。所有细节虽简单却精益求精。此项目的成本没有超过波兰的传统房子。

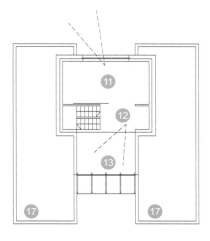

SECOND FLOOR

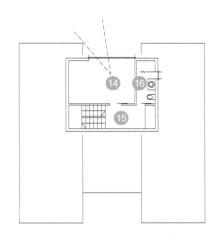

TOP FLOOR

ELEVATION NORTH

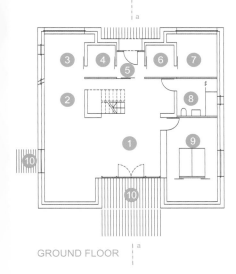

GROUND FLOOR

1   living room
2   dining room
3   kitchen
4   storage
5   porch
6   wardrobe
7   laundry
8   bath
9   sleeping room
10 terrace
11 studio
12 gallery
13 patio
14 room with view
    onto the lake
15 corridor
16 bath
17 green roof
18 photovoltaics
19 solar panels
20 winter garden

SECTION a-a

ELEVATION WEST

SCALE 1:200

0 1   5 m

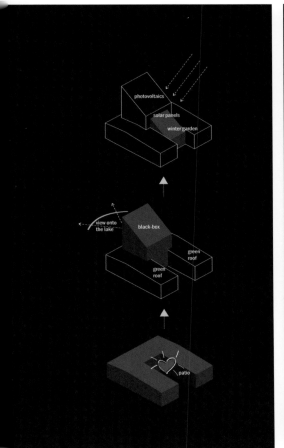

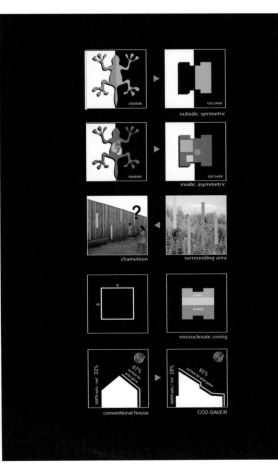

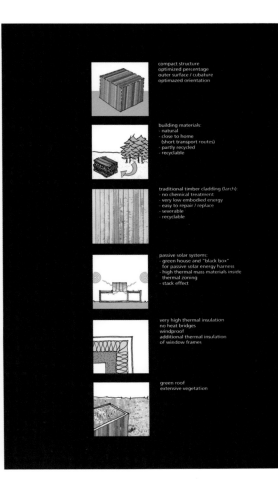

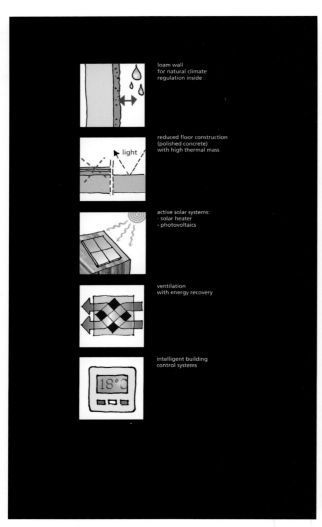

loam wall
for natural climate
regulation inside

reduced floor construction
(polished concrete)
with high thermal mass

light

active solar systems:
- solar heater
- photovoltaics

ventilation
with energy recovery

intelligent building
control systems

18°C

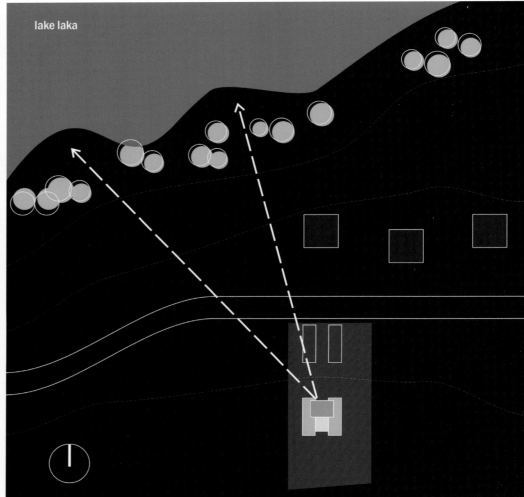

lake laka

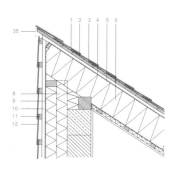

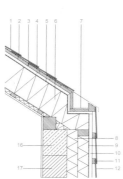

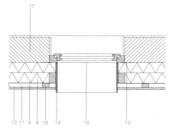

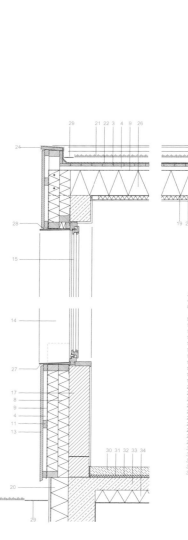

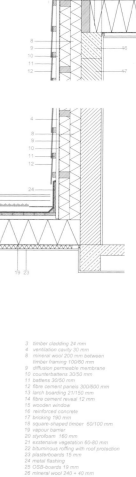

Roof

1 fibre cement panels 300/600 mm
2 roof paper
3 timber cladding 24 mm
4 ventilation cavity 30 mm
5 diffusion permeable membrane
6 mineral wool 260 + 40 mm
7 zink metal box gutter
8 mineral wool 200 mm between
   timber framing 100/50 mm
9 diffusion permeable membrane
10 counterbattens 30/50 mm
11 battens 30/50 mm
12 fibre cement panels 300/600 mm
13 larch boarding 21/150 mm
14 fibre cement reveal 12 mm
15 wooden window
16 reinforced concrete
17 bricking 190 mm
18 square-shaped timber 60/100 mm
19 vapour barrier
35 metal flashing

Reveal

1 fibre cement panels 300/600 mm
2 roof paper
3 timber cladding 24 mm
4 ventilation cavity 30 mm
5 diffusion permeable membrane
6 mineral wool 260 + 40 mm
7 zink metal box gutter
8 mineral wool 200 mm between
   timber framing 100/50 mm
9 diffusion permeable membrane
10 counterbattens 30/50 mm
11 battens 30/50 mm
12 fibre cement panels 300/600 mm
13 larch boarding 21/150 mm
14 fibre cement reveal 12 mm
15 wooden window
16 reinforced concrete
17 bricking 190 mm
18 square-shaped timber 60/100 mm
19 vapour barrier

3 timber cladding 24 mm
4 ventilation cavity 30 mm
8 mineral wool 200 mm between
   timber framing 100/50 mm
9 diffusion permeable membrane
10 counterbattens 30/50 mm
11 battens 30/50 mm
12 fibre cement panels 300/600 mm
13 larch boarding 21/150 mm
14 fibre cement reveal 12 mm
15 wooden window
16 reinforced concrete
17 bricking 190 mm
18 square-shaped timber 60/100 mm
19 vapour barrier
20 styrofoam 160 mm
21 extensive vegetation 60-80 mm
22 bituminous roffing with root protection
23 plasterboards 15 mm
24 metal flashing
25 OSB-boards 19 mm
26 mineral wool 240 + 40 mm
27 drainage area
28 silicone
29 grit
30 polished floor pavement 80 mm
31 styrofoam 20 mm
32 waterproofing
33 concrete slab 120 mm
34 styrofoam 120 mm

CO2-SAVER , POLAND
ARCHITECT: KUCZIA

SCALE 1:20

0   1 m

Zuber 房子

Zuber House

The design of the house was inspired by the place and its traditional architecture. Economical layout and compact shape of the building save resources, energy and money. The structure is divided into two blocks: one with a double pitch roof which is clad in fibre cement and another – used temporarily – with a flat roof and timber facade. The house utilizes passive and active solar components, green materials a technologies, including recycled materials, solar panels and high performance insulat systems. A green roof planted with drough-resistent plants cooperates in reducing h transfer through the roof. The house was built on a very limited budget.

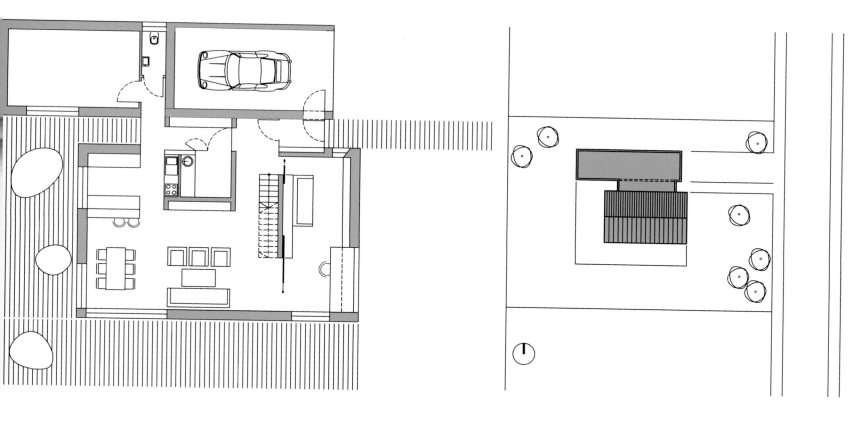

房子的设计灵感来自其所在位置和传统建筑。建筑经济型的布局和紧凑的形状节约资源、能源和金钱。其结构被分为两部分：一部分是纤维水泥覆盖的双坡屋顶，另一部是暂时使用的木制立面和平屋顶。该房子使用了诱导式和主导式太阳能组件、绿色材料和技术，包括可回收再利用材料、太阳能电池板和高性能隔热系统。种植了耐干旱植物的绿色屋顶减少通过屋顶传入的热量。房子是在一个非常有限的预算中建成的。

布面楼房

Fabric Facade Almere

The house is built on one of the 350 plots designated for construction by private owners at Almere near Amsterdam in the Netherlands. Private building is, in the Netherlands with its high population density, not extremely common as land is scarce.

These "Herenhuis" plots, part of the OMA urban plan, had strict envelop requirements. Minimal and maximal heights depending on the position: at the front (min/max 7-14m), middle (4m) or rear (4-7m). The client chose to concentrate his volume in the front part of the plot. Another urban planning requirement was that the ceilings at ground level should be higher than usual (have a clearance of 3.5m) so that the ground floor space is both suitable for residential and work functions.

The client's assignment was: downstairs an artist' studio and exhibition space and the living space on the top floors. The principal had lived in Canada for many years, and expressed the clear wish that the house should not be minimally seized and narrow (like many Dutch residences) but spacious, open and giving one a sense freedom of movement, a dream assignment for space loving designers, with the added challenge of a very limited budget.

The sense of space is created by making a continuous connection between all rooms of the house. The artist's studio downstairs is connected to the living room on the first floor by one of the atriums and the ground floor thereby also gets more light deep into the studio space through this extra 1st floor window.

In the middle of the building volume a central atrium was carved out, with at the top a skylight, visually connecting the entire upper floors to the living space and creates a good light quality and feeling of freedom. Additional advantage is that the study / TV room, top level street side, did not then need windows nor received them to save costs. Due to the split-level in the front part of the living room the ceiling height reaches 4.5m giving it a real mansion ("herenhuis") quality. The dimensions of the doors are 3.5m width x 2.7m height and this also contributes to this sense of openness. But due to all those large measurements the scale of the building is quite difficult to read and can only be measured by comparing it with the traditional sizes of the neighbouring houses.

该楼房建在荷兰阿姆斯特丹附近的阿尔梅勒私人业主指定用于建设的 350 个地块之一上。由于土地稀缺，私人楼房在人口密度高的荷兰是极为罕见的。

这些是 "Herenhuis" 地块 OMA 城市规划的一部分，有着严格的覆盖要求，即根据位置确定的最小和最大高度如下：前部（最小 / 最大为 7~14m），中部（4m）或后面（4~7m）。委托人选择侧重地块前部的数量。城市规划的另外一个要求是，地面层的天花板应该比平常的高（间隔为 3.5m），以便第一层面空间可同时具备居住和工作的功能。

委托人给出的任务是：楼下是艺术工作室及展览空间，顶层是居住空间。业主曾在加拿大居住多年，明确表达了他的愿望，即这个房子的最低限度不能被占用且狭长的（像许多荷兰住宅那样），而应该是宽敞、开放，给人一种运动感觉的自由;

这是热爱空间设计师的梦想。还有一个挑战就是，要在非常有限的预算下完成。

空间感是通过连接所有房间之间的连续性创建。楼下的艺术工作室通过其中一个天井连接到一楼客厅，地面层工作室的采光从而也得到加强。

建筑中间是一个中空天井，顶部有一个天窗，从视觉上连接整个楼上空间和生活空间，营造了良好的光线和自由的感觉。另一个的优点是，为了节省成本，顶层街道边的书房 / 电视房不需要窗户。由于前部客厅的天花板高度为 4.5m，给它一个真正的豪宅质量。门的尺寸为 3.5m 宽 ×2.7m 高，也有助于展示其开放性。但由于这些大型建筑物的规模是很难识别，只能通过邻近房子大小的传统尺寸来比较。

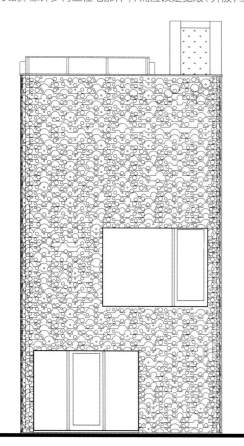

west facade – street

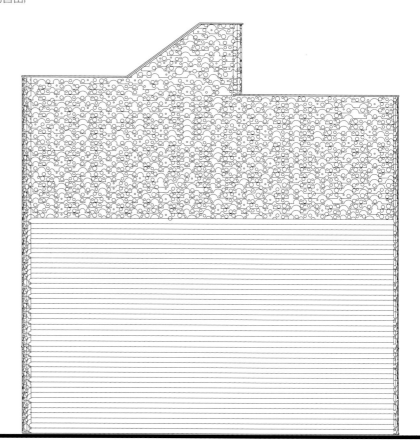

south facade

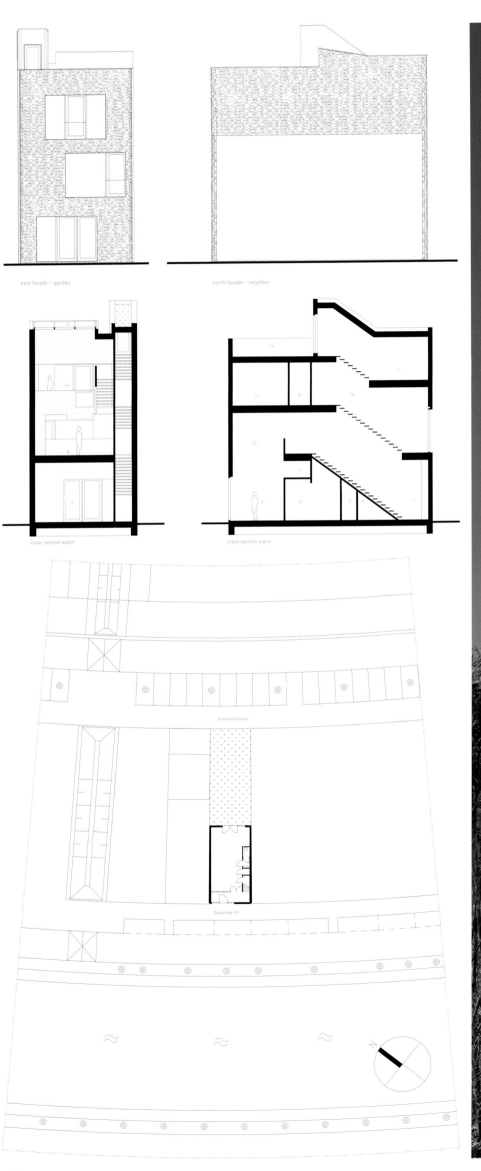

east facade - garden

north facade - neighbor

cross section width

cross section stairs

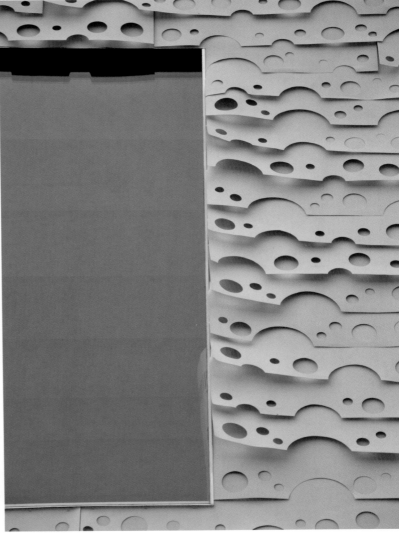

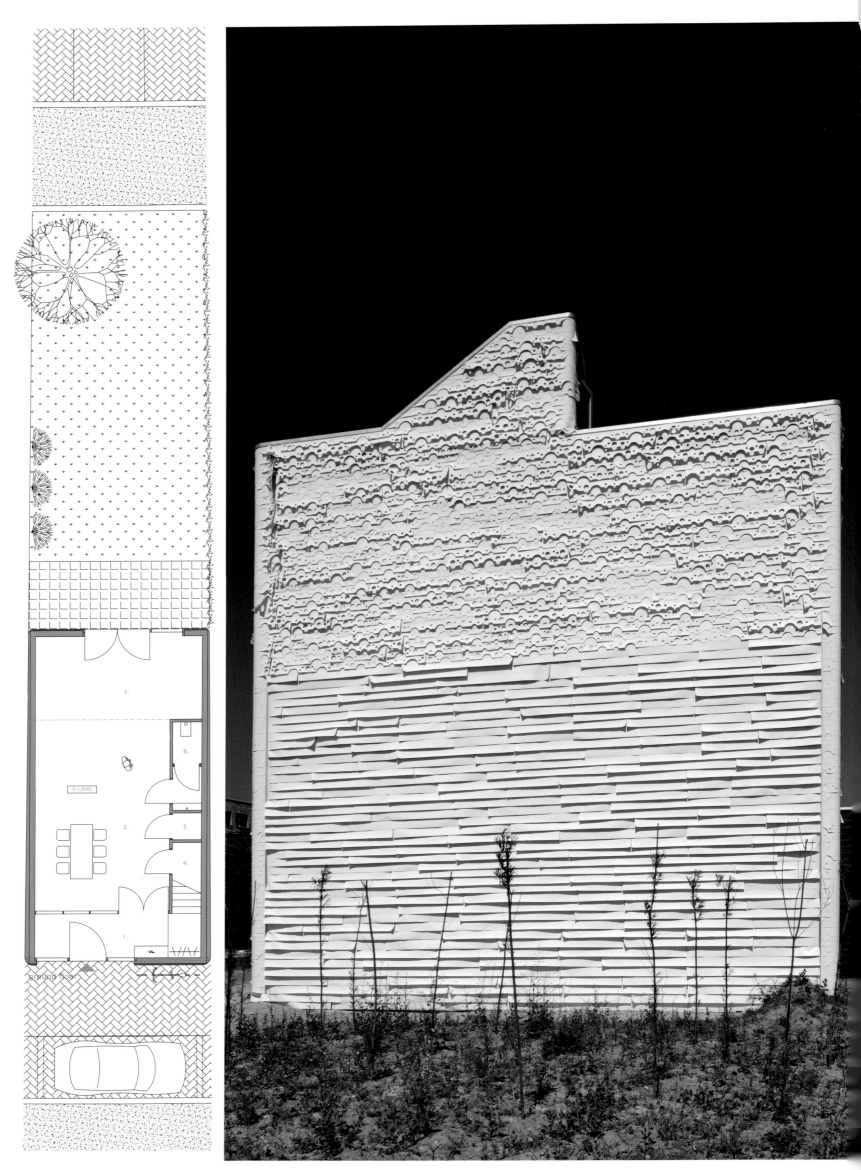

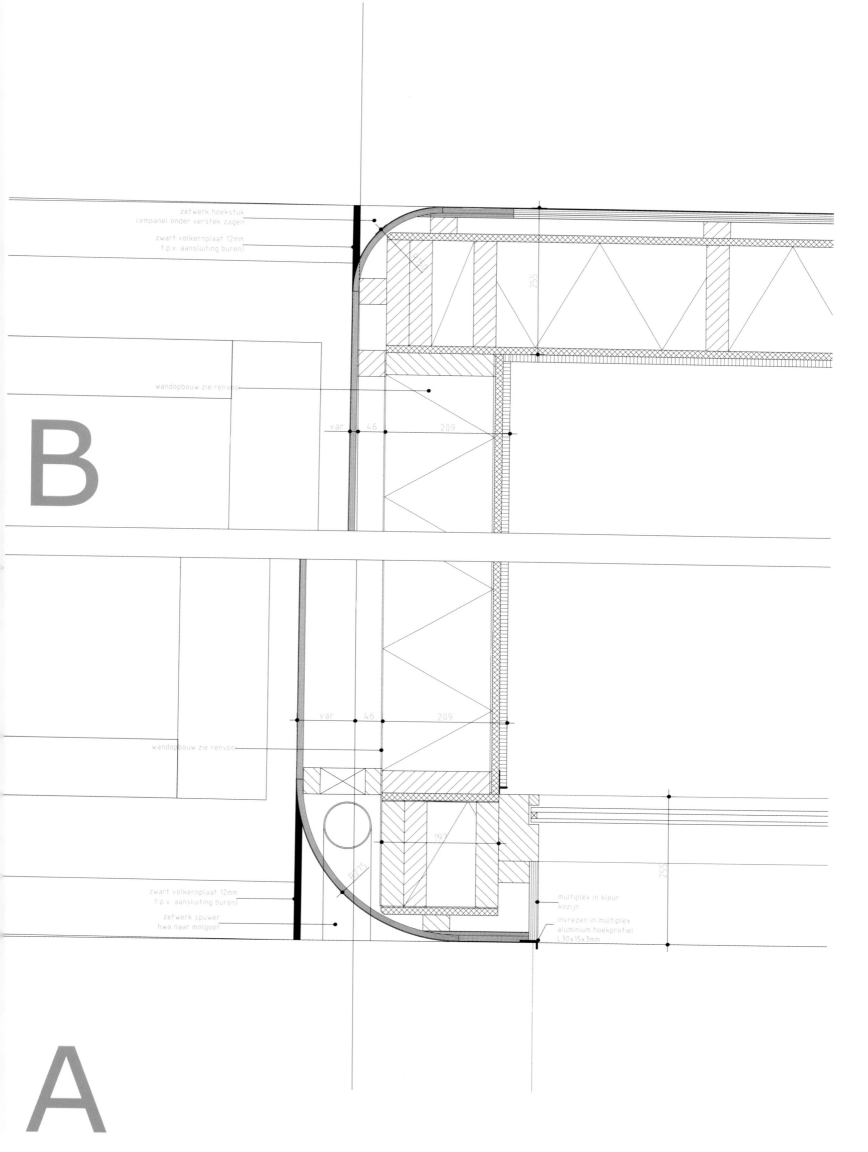

zetwerk hoekstuk
cempanel onder verstek zagen

zwart volkernplaat 12mm
t.p.v. aansluiting buren]

wandopbouw zie renvooi

B

var   46   209

var   46   209

wandopbouw zie renvooi

197

multiplex in kleur
kozijn

zwart volkernplaat 12mm
t.p.v. aansluiting buren]

zetwerk spuwer
hwa naar molgoot

invrezen in multiplex
aluminium hoekprofiel
L 30x15x3mm

A

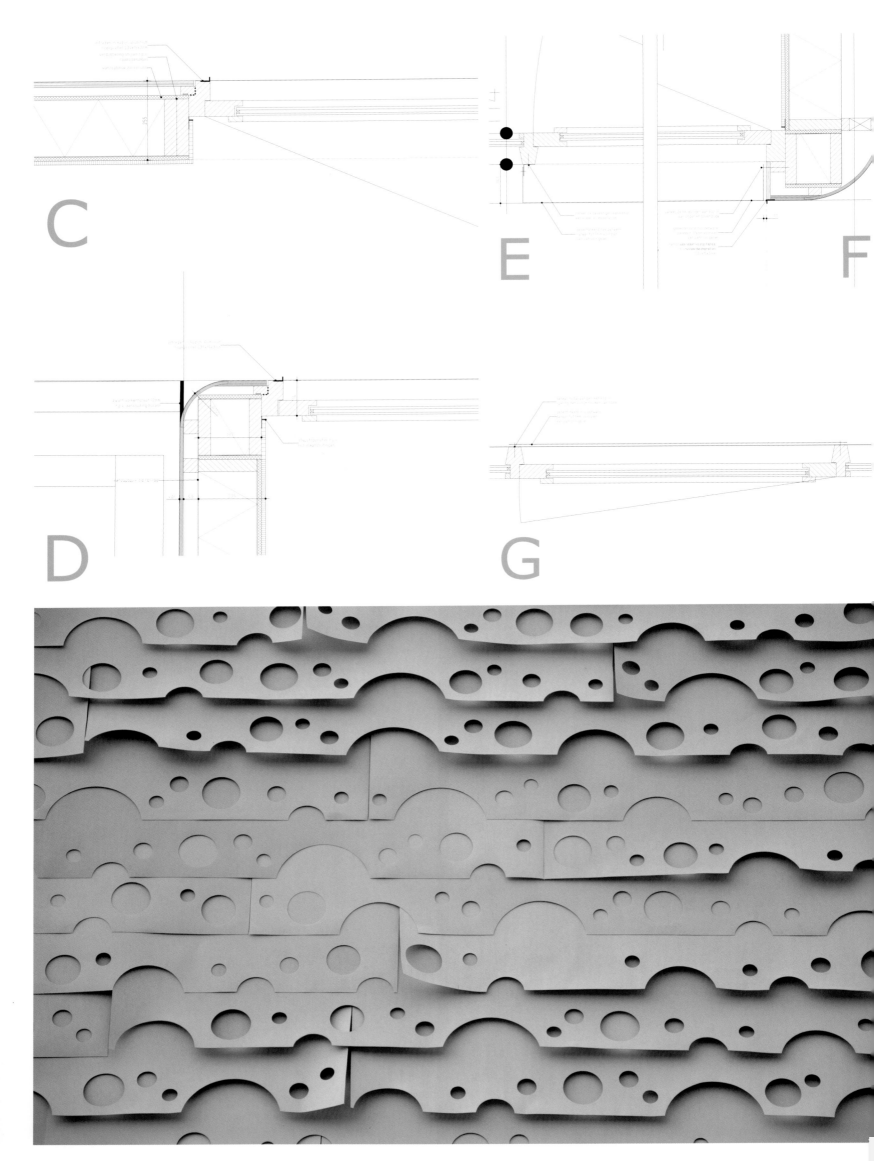

C

E

F

D

G

3

paneel bevestigen aan verzinkte stalen bovenregel
70x15mm, rubberen afstandhouders toepassen
balustrade bevestigen aan kozijn

1000 +vloer
bk. afwerkvloer

geperforeerd rvs zetwerk
paneel t=1.5mm voorzien
van patroon gevel

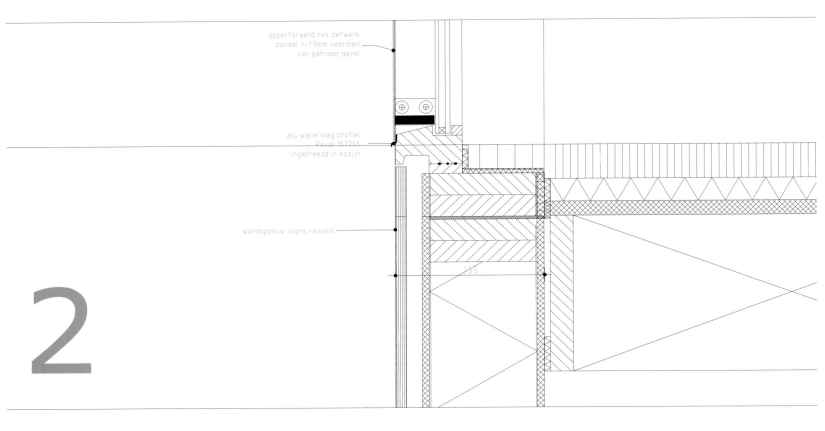

geperforeerd rvs zetwerk
paneel t=1.5mm voorzien
van patroon gevel

alu waterslag profiel
Reval IS7265
ingefreesd in kozijn

2

wandopbouw vlgns renvooi

255

wandopbouw vlgns renvooi

255

aluminium zetwerk lekdorpel
rondom woning toepassen

muisdicht rooster op ventilatiekoker

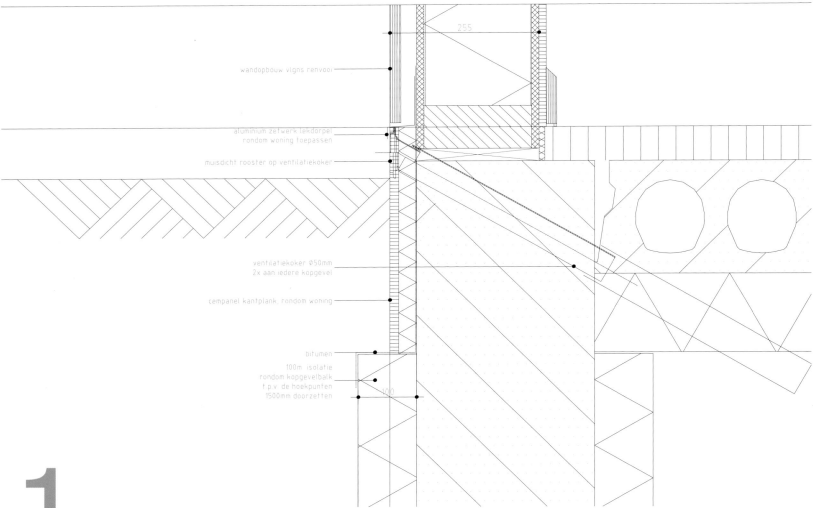

ventilatiekoker Ø50mm
2x aan iedere kopgevel

cempanel kantplank, rondom woning

bitumen

100m isolatie
rondom kopgevelbalk
t.p.v. de hoekpunten
1500mm doorzetten

100

1

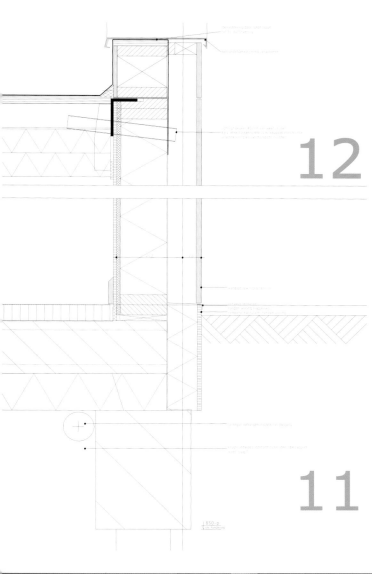

12

11

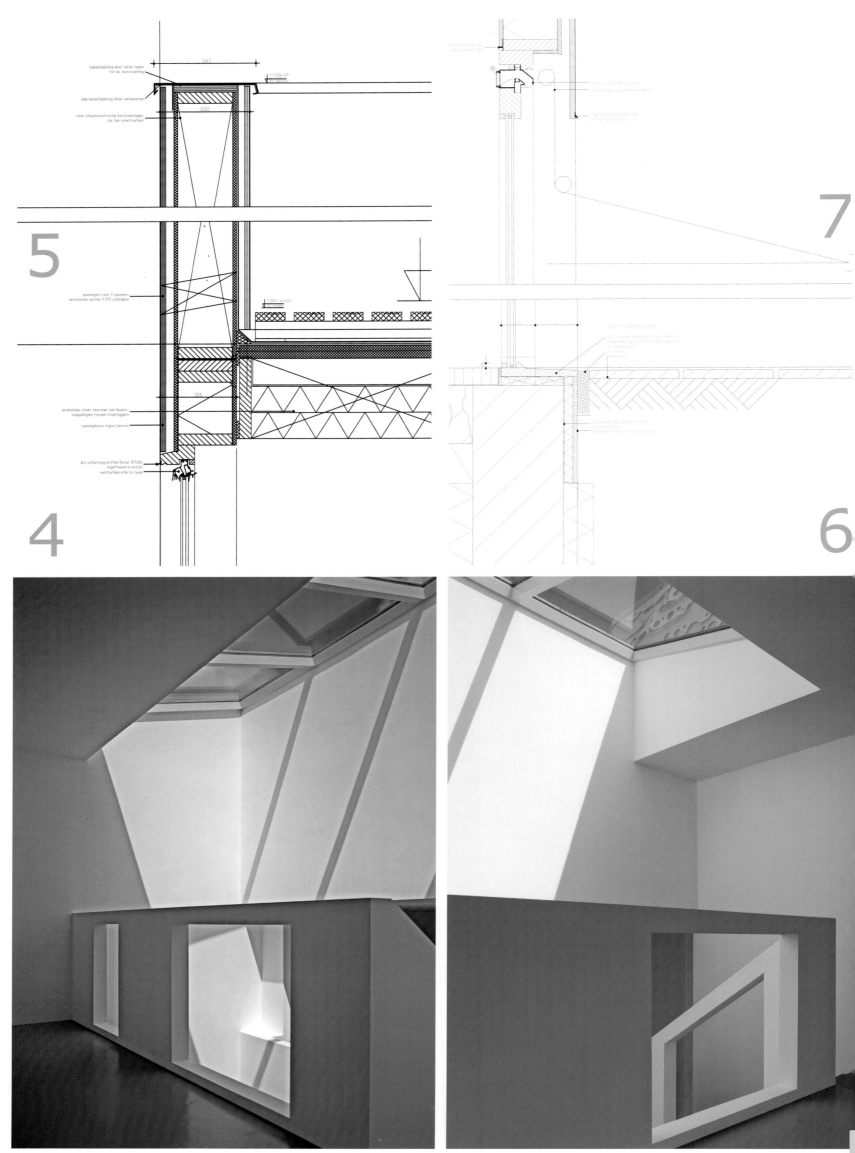

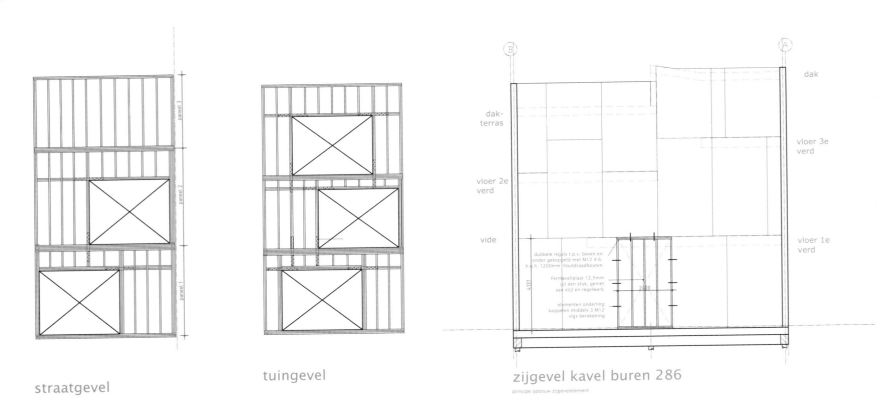

straatgevel

tuingevel

zijgevel kavel buren 286

pirncipe opbouw zijgevelelement

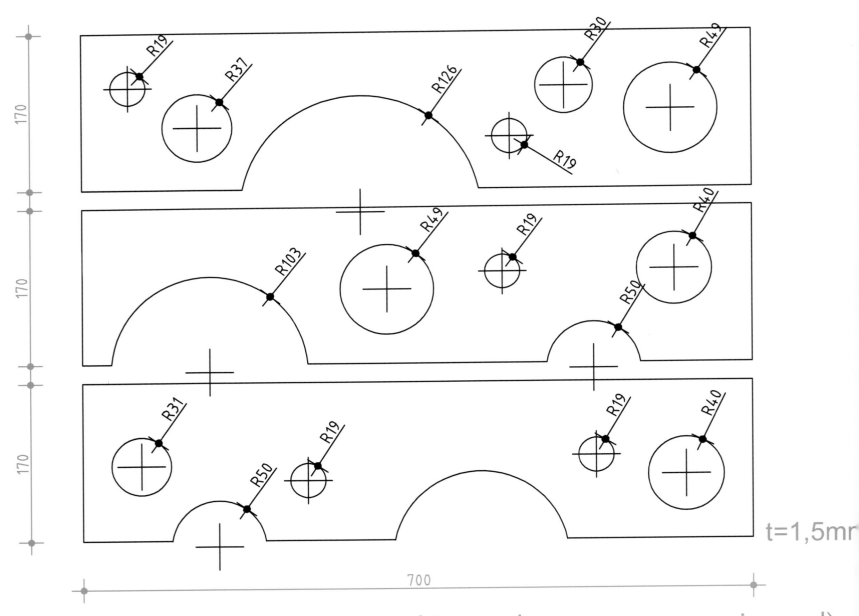

patterning templates (used in random sequence+mirrored)

山屋
Hill House

The Problem / Opportunity

Design is complex. It is more complex to design a home, however fundamental issues offer an architect a starting point: where is the sun? How do we capture it in winter and how do we exclude it in summer?

Original Conditions

The site faces north therefore relegating the backyard, the family's primary outdoor space, to shadow throughout the year. In the 1990s a two-storeyed extension was added reducing solar access even further while creating deep dark space within the house.

Response

Rather than repeating past mistakes and extending from the rear in a new configuration, the proposal was to build a new structure on the rear boundary, the southern edge of the block. The new structure faces the sun employing passive solar gain, saturating itself with sunlight. The new structure faces the original house.

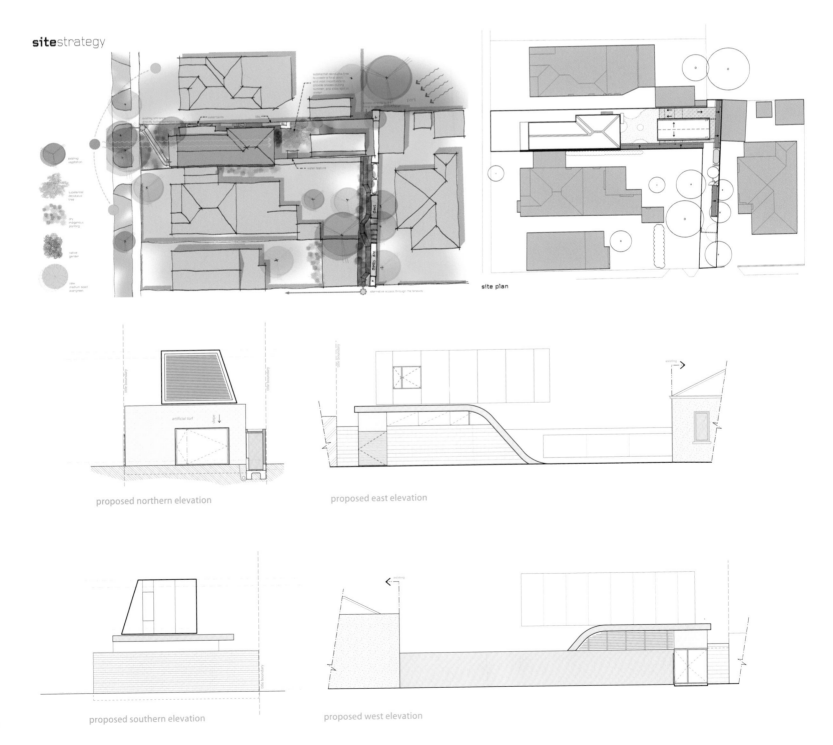

sitestrategy

site plan

proposed northern elevation

proposed east elevation

proposed southern elevation

proposed west elevation

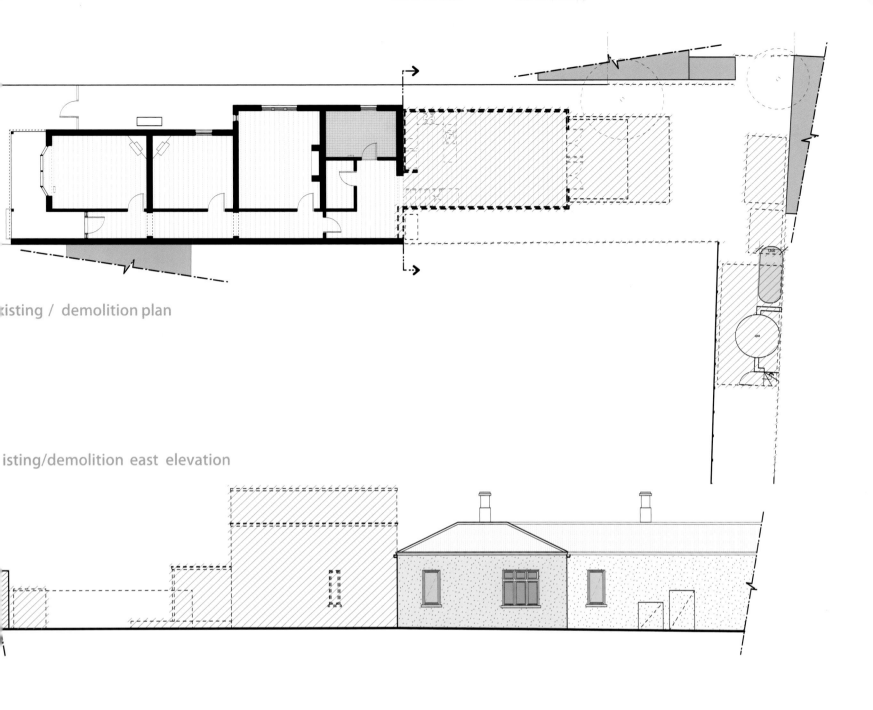

existing / demolition plan

existing/demolition east elevation

Steel

Steel not only provides a solution for the architectural form explored, but also is the primary celebrated material within the small strategic palette applied. The black monolith is a continuous, full height steel truss. The monolith cantilevers more than it is grounded in the hill. The central truss is celebrated in the living space.

The entire load of the second storey travels down the deliberately fragile tri-post in the dining area. The concealed steel posts beside the kitchen counter intuitively tie down the monolith, stopping it from falling forward. The roofing is white colorbond, used strategically to reduce thermal load. The kitchen benches are steel, used deliberately as it is both robust and slowly revealing a beautiful patina of age. The door here pivots, seemingly defying gravity. The stairs, the doors and the windows are all steel. From the large and robust to the fine and detailed, steel has been celebrated as both a structural solution and an aesthetic.

问题 / 机遇

设计是复杂的，住宅设计更是复杂无比，但根本问题是给建筑师提供一个出发点：太阳在哪里？如何在冬天捕捉太阳，如何在夏天避开它？

原情况介绍

该场地朝北，因此把后院，即该家庭的主要户外空间看成全年都是阴影笼罩的地方。在 20 世纪 90 年代扩建的一栋二层建筑在房子内部形成深暗空间的同时进一步减少了太阳的照射。

对策

不再重复过去的错误，也不从后部延伸出一个新的结构，该项目建议在房子后面的边界——南部边缘建一个新的建筑。新建筑面对太阳利用了诱导式太阳能增益。让其沉浸在太阳光中。新建筑跟原房子是面对面的。

钢结构

钢材不仅是裸露建筑形式的材料，也形成了建筑的主色调。黑色的整体是一块连续的全高钢桁架。整个桁架超出基底部分悬挑出来，中央桁架则在居住空间上。

第二层的整个负载慎重地沿着脆弱三柱直达用餐区。厨房柜台上隐蔽的钢柱直观地承托上部桁架防止其下落。盖屋顶的材料使用白色的镀铝锌钢板，以降低热负荷。厨房长凳都是钢制的，因为钢材是坚固的，并能慢慢地揭示岁月的美丽光泽。门在此处旋转，看起来经受得住重力。楼梯、门窗都是钢制的。从大型强壮到精细细节，钢材从结构性和美观性都在此建筑中表露无遗。

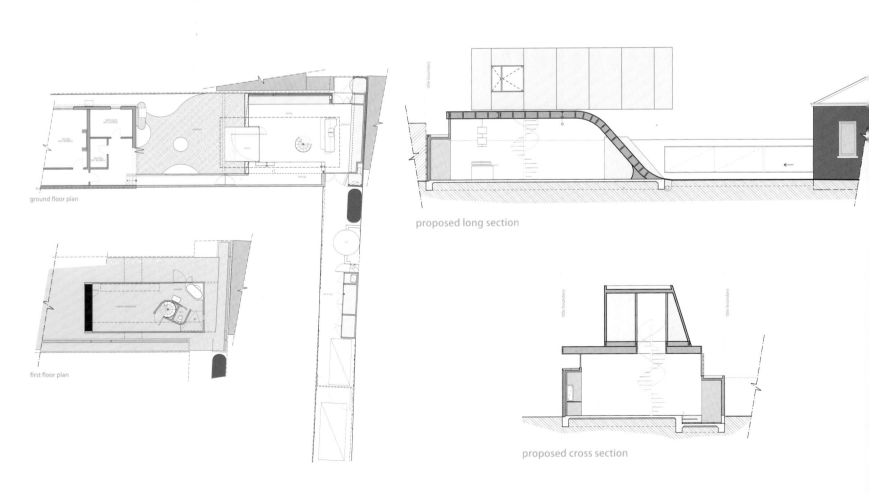

ground floor plan

first floor plan

proposed long section

proposed cross section

Sustainability

Hill House is far more sustainable than first appearance suggests. The whole strategy was to get the house in the backyard to face the sun and to get passive solar performing optimally. All windows are double glazed and LowE coated. The yard is water efficient – the use of synthetic grass with strategically placed garden patches create dense areas of planting, resulting in very little demand for water.

Long strips of windows to the East and West have been equipped with operable louvers. The north-facing facade consists of an entire wall of the same mechanically-operated louvers, providing the option of controlling cross-winds and sunlight. The grass on the hill envelops the ground floor in an additional layer of insulation, it is a thermal roof blanket, installed to supplement the existing insulation of the building structure beneath whilst also protecting the roof. A white roof is used throughout to increase solar reflectance, sustainably reducing heat gain within the house.

可持续发展性

　　山屋的可持续性比其外表显示出来的要多得多。整个设计策略是让后院的房子正对太阳，将诱导式太阳能进行优化。所有窗户都是双层玻璃和LowE（低辐射）涂层。院子采用节水技术——人造草坪的使用与有策略地设置园地形成植被密集区，结果是对水的需求变得非常小。

　　东、西向的长条窗户都配有可手动调节的百叶窗。北立面的整面墙都是由相同机械操作的百叶窗构成，可控制侧风和阳光。山上的草覆盖地面作为绝缘附加层，是一种安装在屋顶的热毯，补充建筑现有的隔热层，在保护膜顶的同时也保护屋顶建筑结构。整栋建筑都是用白色屋顶，目的是提高太阳能反射率，持续减少房子里的热增益。

巴特勒之家

Butler House

The vertical and architectural pinnacle of the Butler House, fills the void that affects so many inner — city dwellings-a lack of outdoor space. A very open, vertical path of stairs allowed sounds to travel to all corners of the dwelling – the path was helped by ubiquitous reflective surfaces in steel and concrete. The challenge was to reduce sound transmission, but not to a point of isolation – the family still very much enjoyed the connection allowed by cross-level communication. The solution was found in celebrating the very thing that made this house different – it's vertical nature.

An adjustable vertical spine was fashioned from floor to roof, creating a flexible isolation between levels. Operable louvers allows controllable degrees of isolation; timber shelves, which in time will fill with miscellaneous objects, offers further resistance; a strip of lusciously-green carpet flanking the spine introduces a much-needed soft surface. A stack effect was also fostered, where strategic ventilation allowed purging of hot air to the terrace during the summer months.

The existing roof structure was simply cut at the collar-tie and refashioned in a manner, minimizing steel use, to allow a bed for the roof pod – a bionic upgrade of sorts. Sitting the pod within the roofline was imperative in maintaining engagement with the house below. This way, those cooking snacks on the terrace can converse with those preparing salad in the kitchen.

巴特勒之家的垂直和建筑尖顶填补了影响如此多城市住宅的空隙——缺乏户外空间。一个非常开放的、垂直的楼梯让声音传至住宅的所有角落，该楼梯由反射面的钢筋混凝土支撑。设计的挑战是降低声音传播，而不是隔离声音——因为这个家庭仍然非常喜欢跨层交流的联系方式，解决的办法就是这座房子的与众不同之处——垂直。

一根可调节的垂直柱连接天花板和地板，形成了层与层之间的灵活隔离。活动百叶窗调节隔离范围；木架，装满杂物后将提供进一步的隔离；垂直柱侧边的植物绿毯营造了所需的柔软外观。烟囱效应也由此形成，在夏季，重要的通风方式可将热空气排放到屋顶。

现有屋顶结构只是采用在顶部削减和改造的方式，最大限度地减少钢材的使用，让屋顶吊舱房可以放下一张床——类似一种仿生升级。坐在屋顶吊舱内，必须跟下面的房子保持接触。这样的话，在屋顶准备点心的人可以和在厨房里准备沙拉的人聊天。

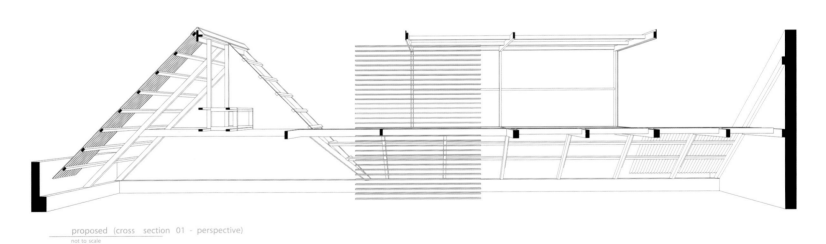

proposed (cross section 01 - perspective)
not to scale

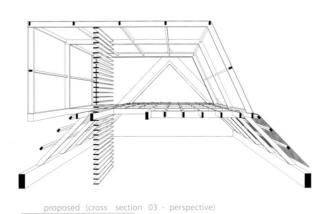

proposed (cross section 03 - perspective)
not to scale

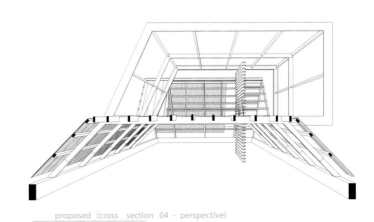

proposed (cross section 04 - perspective)
not to scale

A traffic-able glass floor on the roof, next to the spiral staircase, allows a visual connection and solar access to the living areas below. The effect of the newly-penetrated light throughout the three-storey apartment cannot be downplayed. Again, the adjustable louvers and blind help control the light admittance.

The existing mezzanine floor was used primarily as a rumpus room for the children, it's openness was at times detrimental. The challenge was to cordon off the space to use as a bedroom, without obstructing the all-important windows. The solution came in covering the gap between wall and floor in playful joinery, whilst a cavity to allow light and ventilation to the living room below.

The kitchen functioned-so it was kept. New overhead joinery was added for further storage and as a point of connection to the living space beyond. Incorporated was the new spiral staircase from above, tying it into the architecture of the space. The kitchen, which had previously felt shoved into the corner, was now defined and connected.

The Butler House was a tricky project to approach. The defined nature of the boundaries meant a creative approach had to be adopted in order to make the most of what was available. The result is an adaptable dwelling that will grow and alter over time, just as the family will.

屋顶上可行走的玻璃层，位于螺旋形楼梯的旁边，形成了一个视觉上的联系并让太阳光进入到下面的生活区。可穿透三层公寓的光线效果不能被轻视。再者，活动百叶窗也协助控制光线进入。

现有的中间层原来主要用作儿童娱乐室，它的开放性有时是有害的。面临的挑战是在不阻碍所有重要窗口的情况下，封锁这块空间用作卧室。解决方案是用细木工制品覆盖壁壁和楼层之间的空隙，而洞口允许光线和通风装置通到下面的客厅。

厨房的功能被保留。头顶上新增的细木工制品增加了存储空间并作为与远处生活空间的连接点。在上方并入的新螺旋形楼梯，把它拉入这个空间的建筑。以前觉得被挤到角落的厨房，现在被定义和连接。

巴特勒之家是一个复杂的案例。边界界定意味着必须采用创造性的设计手法，以尽可能利用现有可用的空间。结果一个有适应性的住宅，会随着时间的推移成长和改变，就像家庭一样。

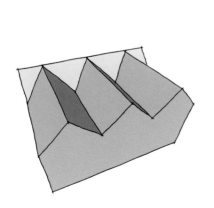
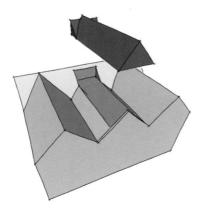
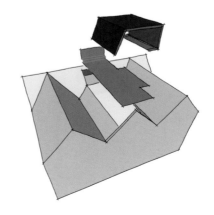
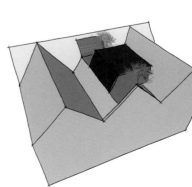

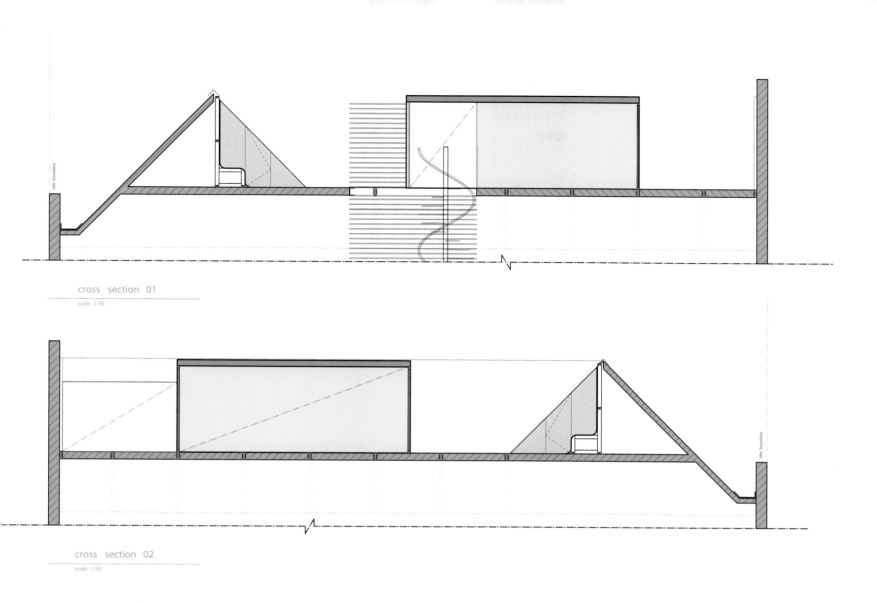

cross section 01
scale 1:50

cross section 02
scale 1:50

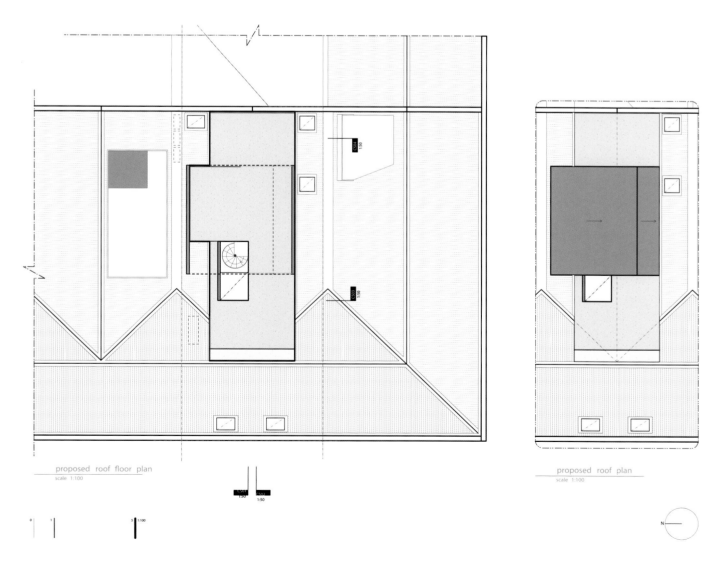

proposed roof floor plan
scale 1:100

proposed roof plan
scale 1:100

N

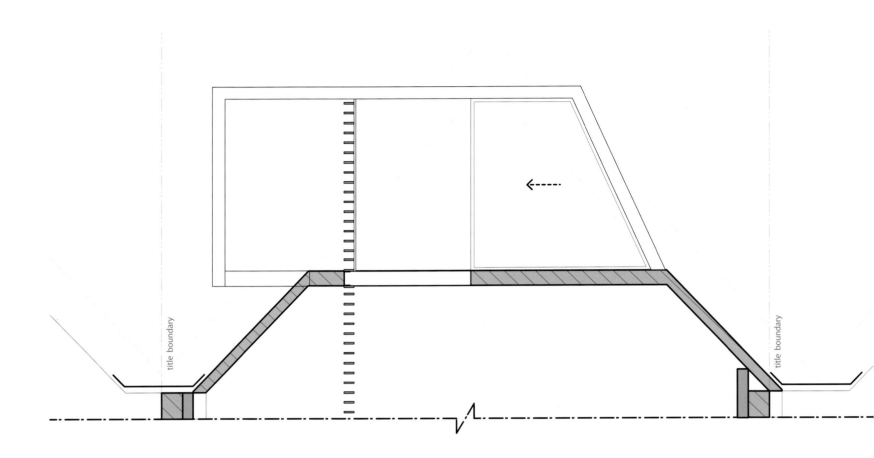

cross section 03
scale 1:50

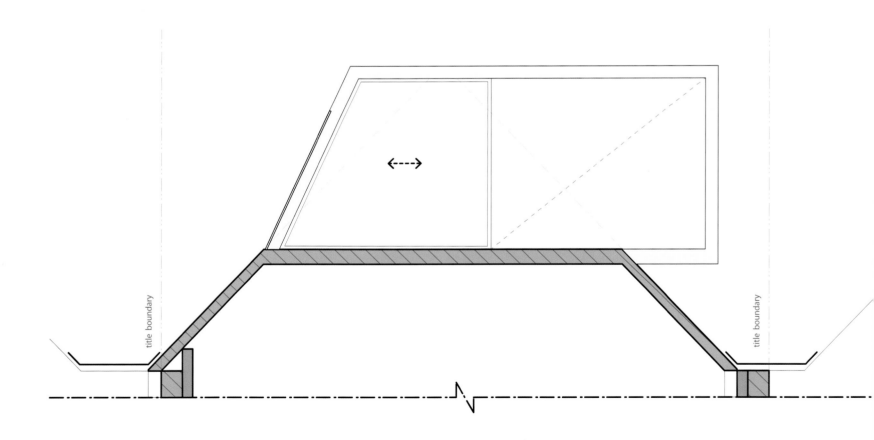

cross section 04
scale 1:50

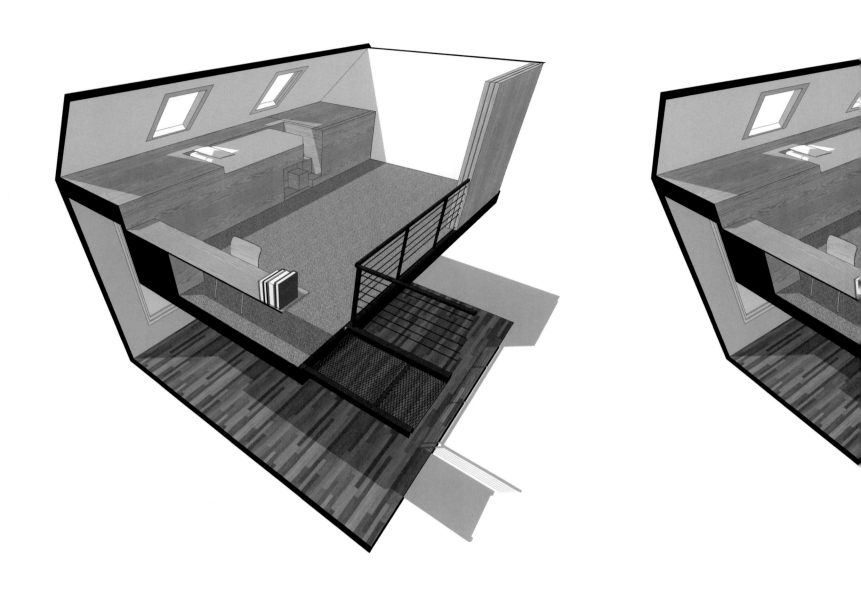

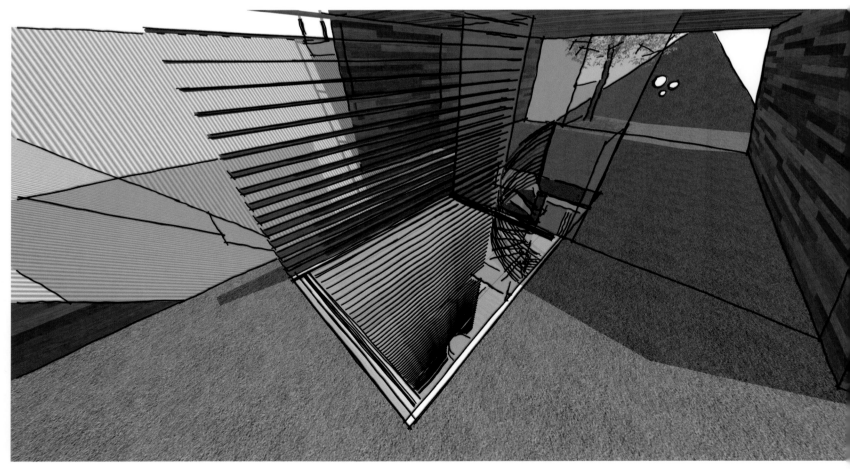

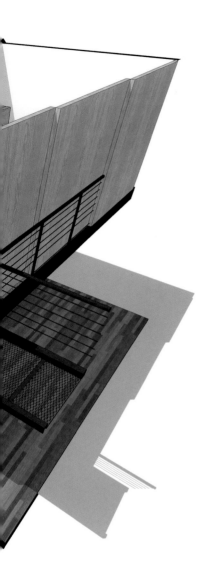

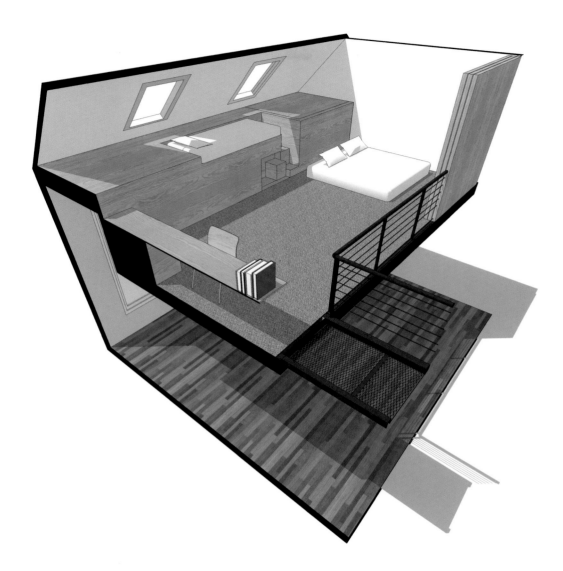

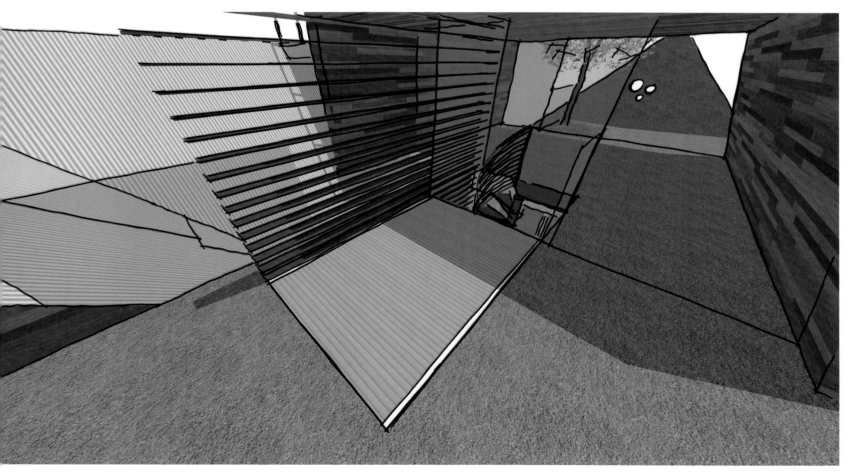

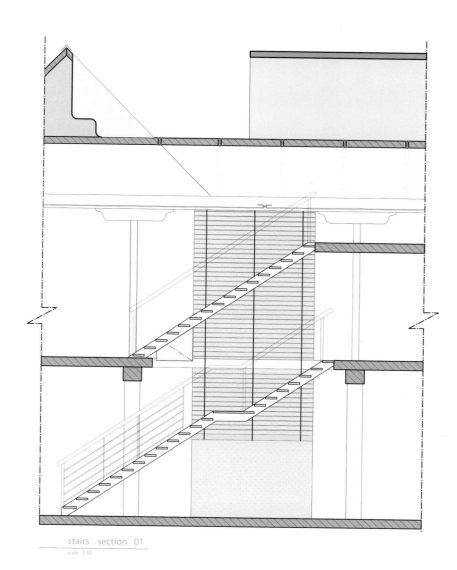

stairs section 01
scale 1:50

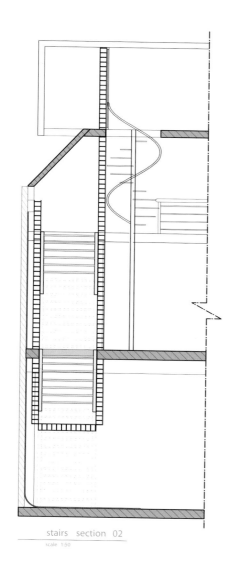

stairs section 02
scale 1:50

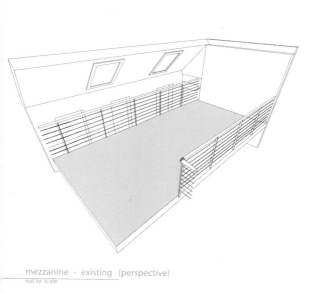

mezzanine - existing (perspective)
not to scale

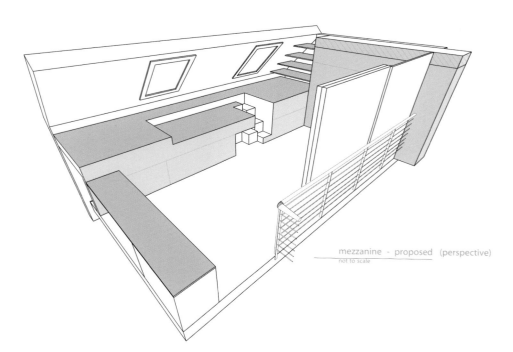

mezzanine - proposed (perspective)
not to scale

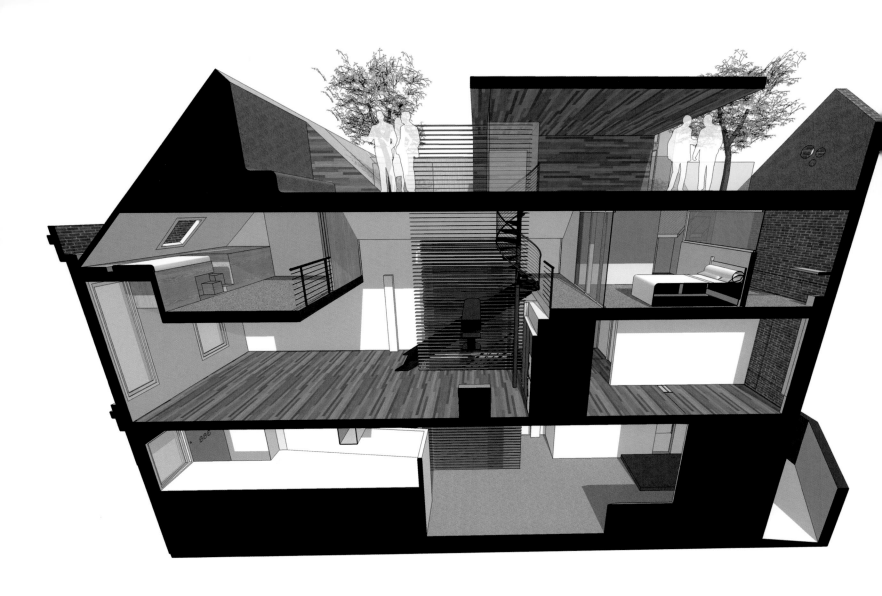

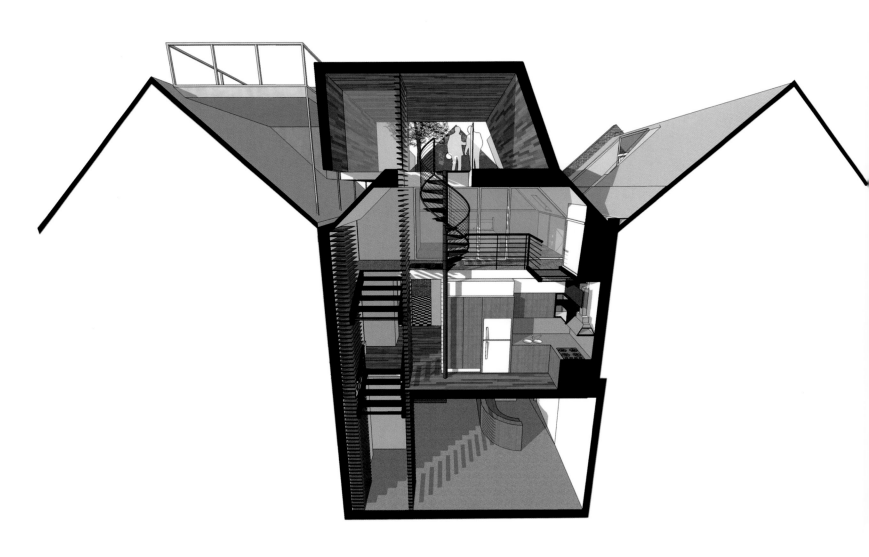

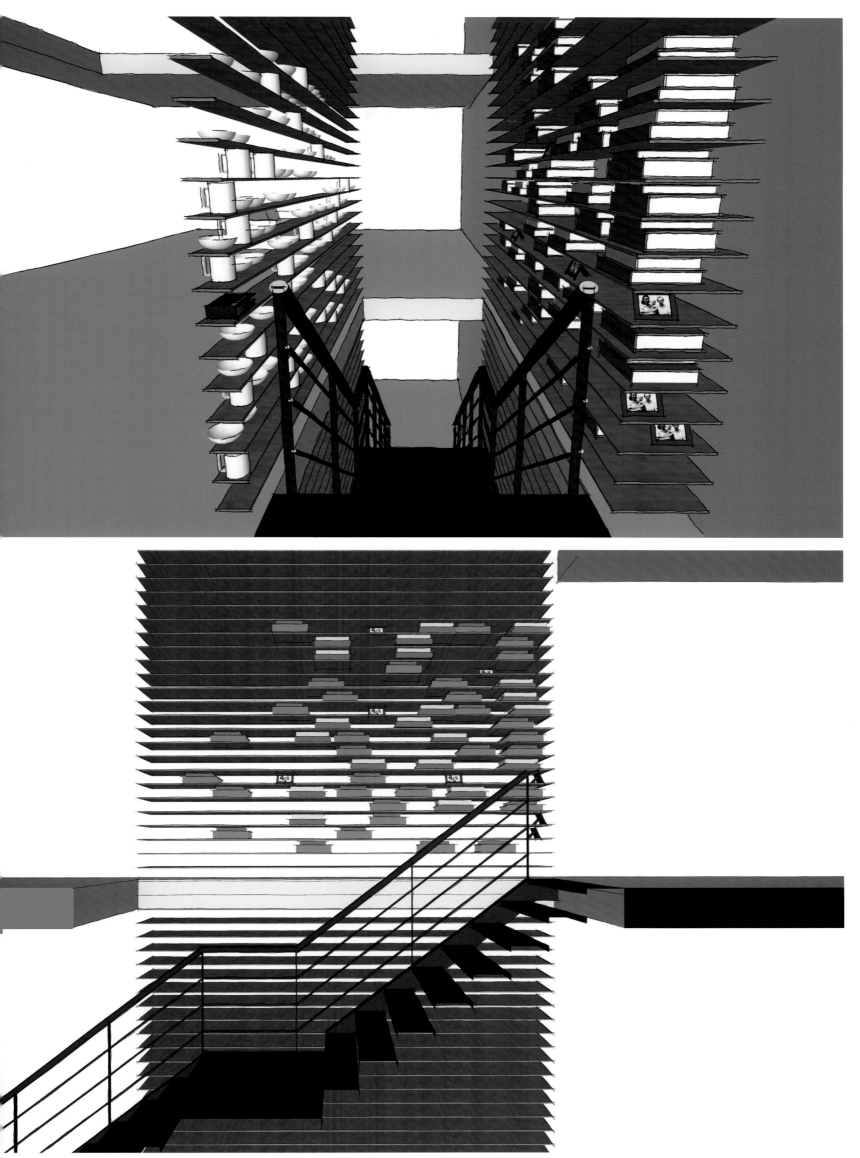

Ilma 林屋

Ilma Grove House

The Ilma Grove House is Andrew Maynard Architect's greenest house (so far). Its planning and orientation is based solidly around passive solar efficiency. All roof captured rain water is harvested. It has solar panels, high performance insulation, recycled materials, LowE coated double glazing, low VOC materials, and most importantly, it is small.

Adhering to the principle of "small is green" (less waste, less electricity consumption, less materials, less cost), the result is a functional open plan, where maximizing passive solar gain becomes indispensable. A master bedroom has been included on the first floor with a roof terrace above that overlooks the city to the south and the Dandenong ranges to the east.

Andrew Maynard Architects took advantage of the north facing backyard, and developed an exploration of mass where segments were carved out in order to maximize sun penetration. This generated a geometrical structure where the internal flesh of the box is revealed with rich timber surfaces, contrasting the raw recycled brickwork.

Ilma 林屋是安德鲁梅纳德建筑事务所设计的最环保的房屋（目前为止），其规划和定位是基于诱导式太阳能效率，所有的屋顶都可以收集雨水。房子有太阳能电池板、高性能绝缘材料、循环再利用材料、LowE（低辐射）双层玻璃和低 VOC（挥发性有机化合物）材料，最重要的是，它很小。

根据"少就是环保（浪费少、耗电少、材料少、成本低）的原则，结果是一个功能性的开放设计，最大化诱导式太阳能增益是不可或缺的。主卧在第一层，屋顶阳台可以俯瞰城市南部和丹德农山脉东部。

安德鲁梅纳德建筑师利用朝北的后院，开展了质量研究，以最大化太阳的穿透力，形成了一个几何结构，林屋的内部实质是用丰富的木材表面展现，与再生原砖形成对比。

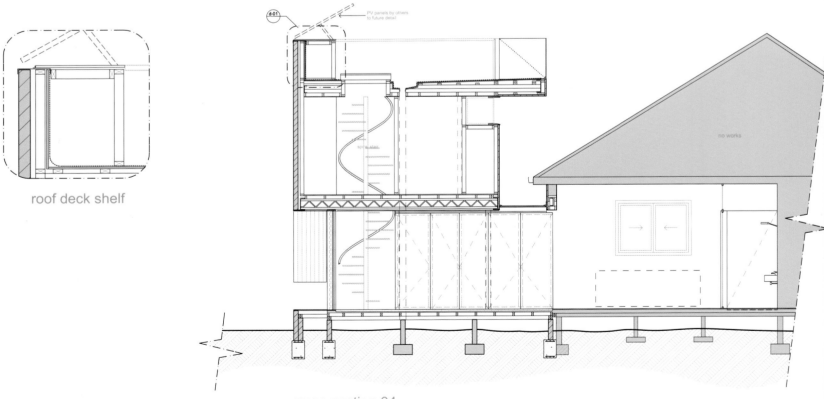

roof deck shelf

cross section 04

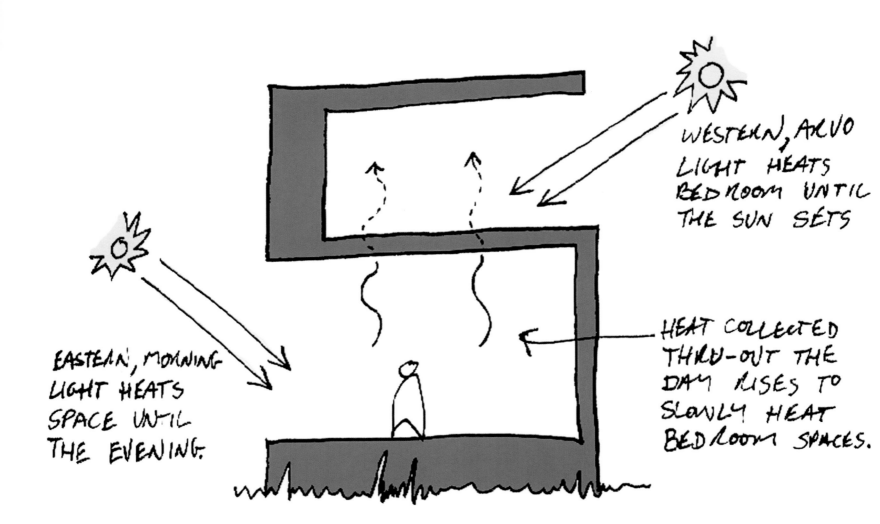

WESTERN, ARVO LIGHT HEATS BEDROOM UNTIL THE SUN SETS

HEAT COLLECTED THRU-OUT THE DAY RISES TO SLOWLY HEAT BEDROOM SPACES.

EASTERN, MORNING LIGHT HEATS SPACE UNTIL THE EVENING.

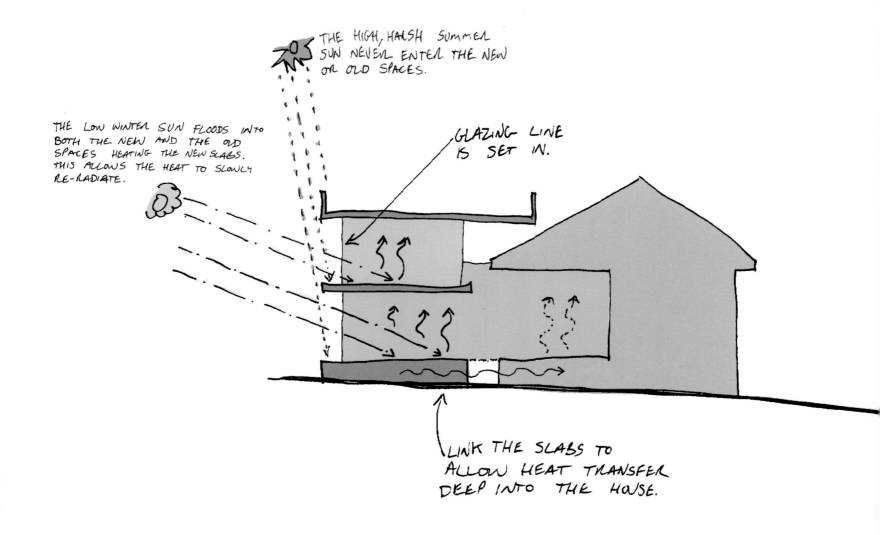

THE HIGH, HARSH SUMMER SUN NEVER ENTER THE NEW OR OLD SPACES.

THE LOW WINTER SUN FLOODS INTO BOTH THE NEW AND THE OLD SPACES HEATING THE NEW SLABS. THIS ALLOWS THE HEAT TO SLOWLY RE-RADIATE.

GLAZING LINE IS SET IN.

LINK THE SLABS TO ALLOW HEAT TRANSFER DEEP INTO THE HOUSE.

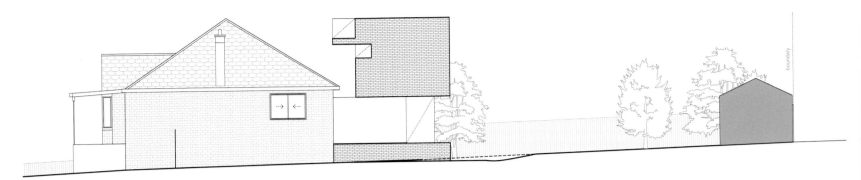

proposed east elevation

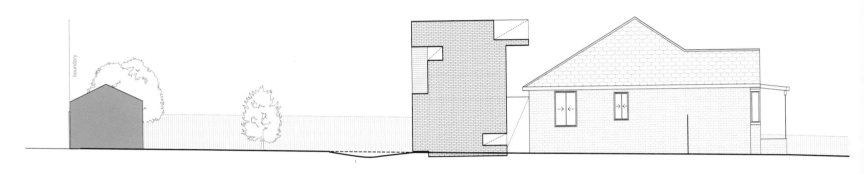

proposed west elevation

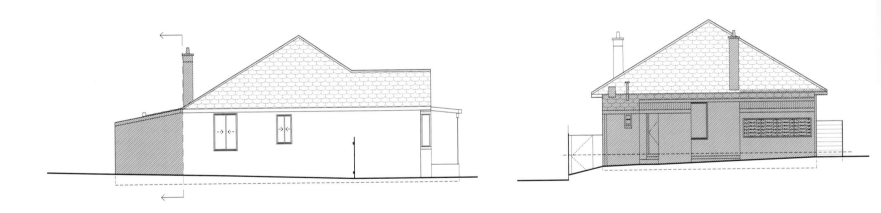

hatch denotes area
to be removed

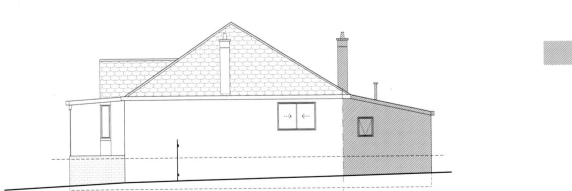

The internal flooring is locally sourced bluestone with the intention that using locally produced materials shortens transport distances, thus reducing CO_2. The dark, dense nature of bluestone acts as a thermal mass soaking up the low winter sun & passively heating the house. In summer the high sun does not come into contact with the bluestone allowing the floor to act as a cooling mass in the hotter months. LowE coated double glazed throughout the house ensures that heat is retained in winter and reduces the penetration of heat in summer. These design features completely remove the need for air-conditioning and drastically reduce the necessity of heating throughout the colder months.

The choice of materials was a vital step in order to create a sustainable structure. The Ilma Grove house reused the bricks of the demolished lean-to on site. This avoids waste, landfill and transportation of materials. Face brick masonry is also a durable and a low maintenance material which can potentially be recycled again. Reinforcing the thermal performance of the recycled brick is high performance insulation that has been installed throughout the home.

After lengthy discussions about the brief the client made a conscious choice not to have an en-suite upstairs. This helped to reduce the size of the addition while also reducing the embodied energy.

Solar panels have been added to make the house coal independent. Ample solar energy is harvested in winter, while a surplus of energy is fed back into the power grid in summer.

Maximizing the outdoor space and connecting with it so that it has become a natural extension of the living space was the key. A small house is a sustainable house.

室内地板是就地取材的青石，目的是缩短材料运输距离，减少二氧化碳排放。深色致密性的青石，充当吸收冬天低温的太阳能和诱导式给房子供暖的热源。夏天，高温的太阳能没有传递到青石，使地板在炎热的季节可充当冷却器。整个房子的 LowE 双层玻璃确保在冬季可防止热量流出和在夏季可阻隔热气进入。这些设计特色完全排除了安装空调的必要，在寒冷的月份大大减少了加热的必要性。

材料的选择是创造可持续性建筑的重要步骤，Ilma 林屋重新利用现场拆除的砖头，避免了材料浪费、填埋和运输。砖头也是一个持久和低成本维护的材料，可被回收再利用。增强回收再利用砖的热性能的是高性能绝缘材料，已在家里全面安装。

经过长时间的讨论，客户决定不在楼上建浴室套间，这有助于减少增建物面积，同时也降低了能耗。

太阳能电池板的使用让房子可以能源独立。在冬季可以收获充足的太阳能，在夏季将剩余的太阳能反馈给电网用于发电。

该项目设计的关键是最大化户外空间并连接使之成为生活空间的一个自然延伸。这个小房子就是一个可持续性的房子。

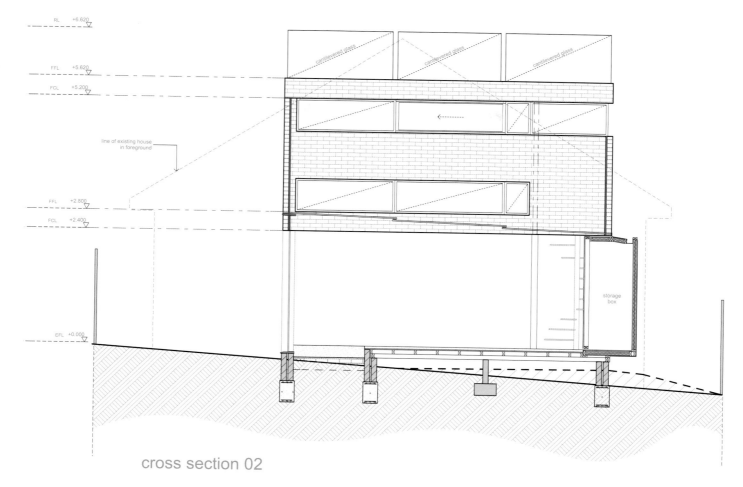

RL +6.620

FFL +5.620

FCL +5.200

line of existing house
in foreground

FFL +2.800

FCL +2.400

EFL +0.000

cantilevered glass

cantilevered glass

cantilevered glass

storage
box

cross section 02

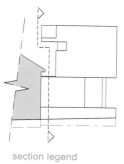

section legend

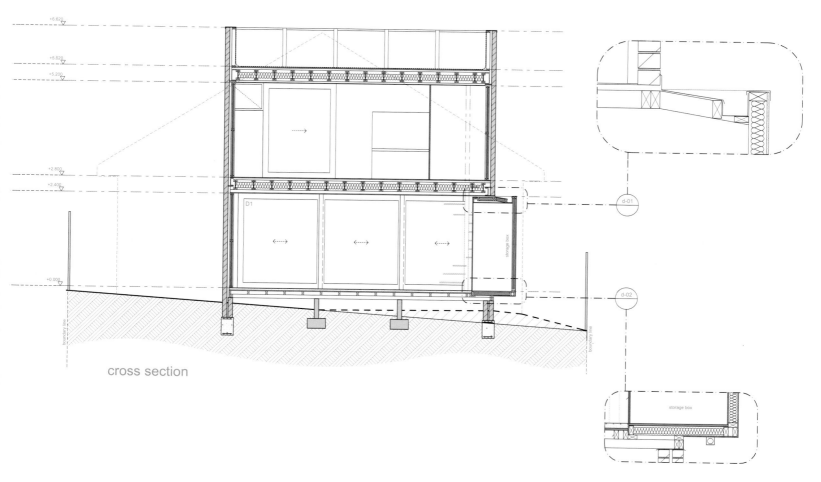

+6.620

+5.620

+5.200

+2.800

+2.400

+0.000

D1

storage box

d-01

d-02

storage box

cross section

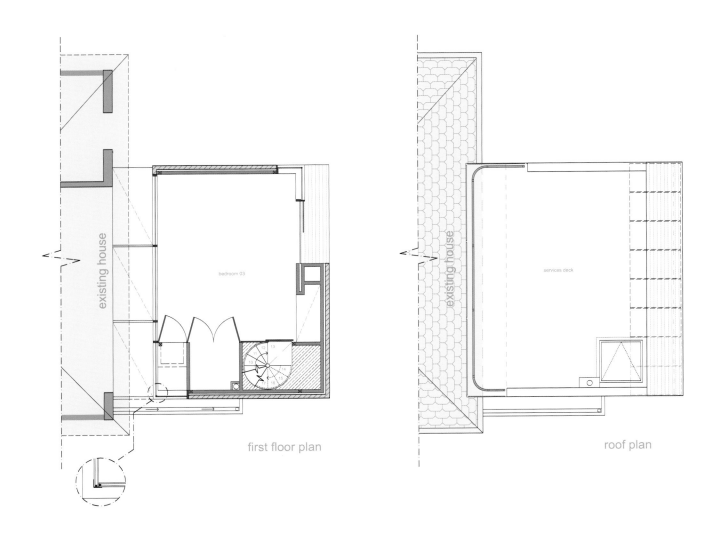

first floor plan

roof plan

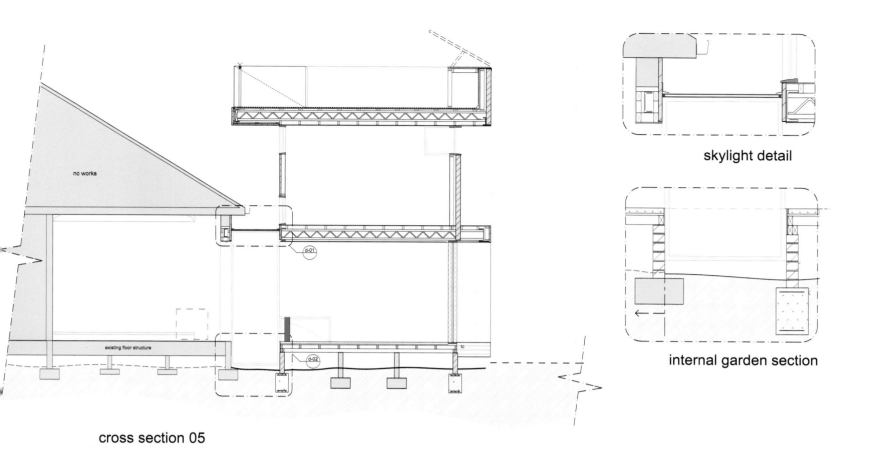

cross section 05

skylight detail

internal garden section

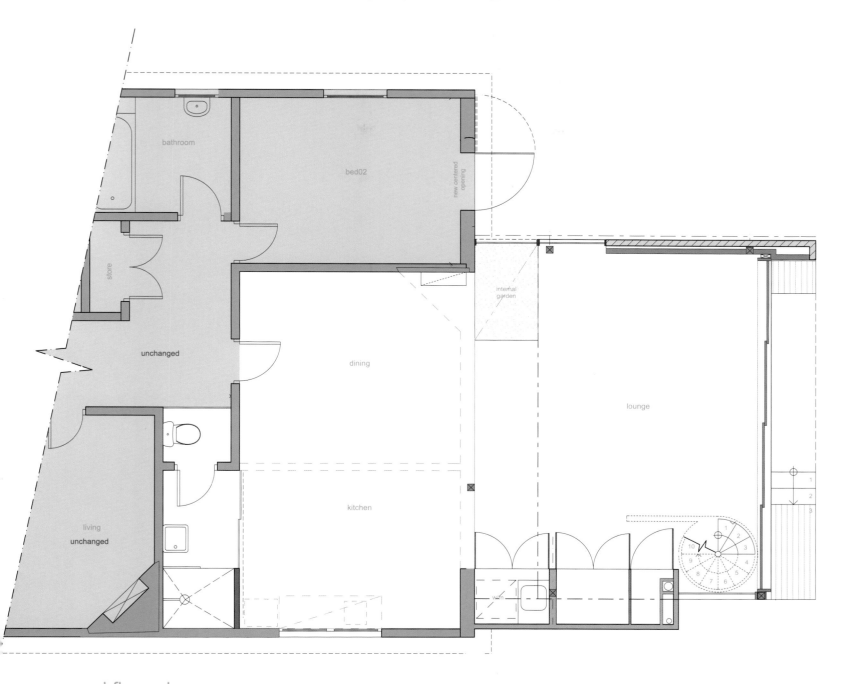

ground floor plan

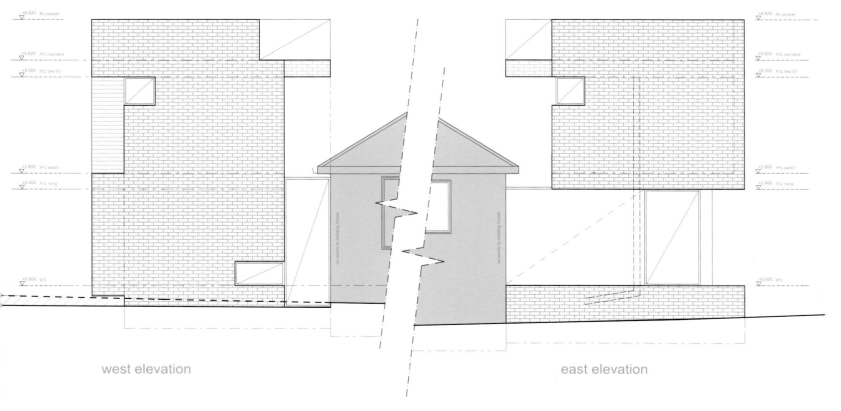

west elevation

east elevation

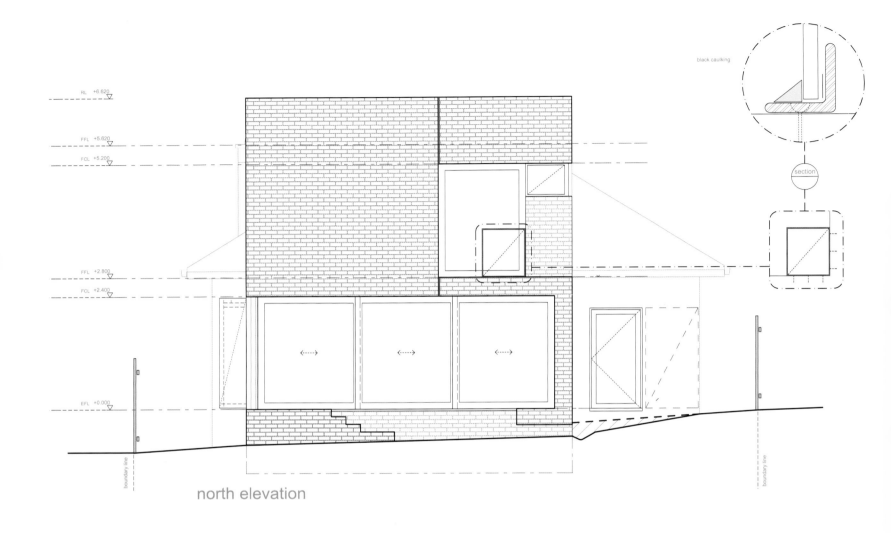

black caulking

section

RL +6.620
FFL +5.620
FCL +5.200
FFL +2.800
FCL +2.400
EFL +0.000

boundary line
boundary line

north elevation

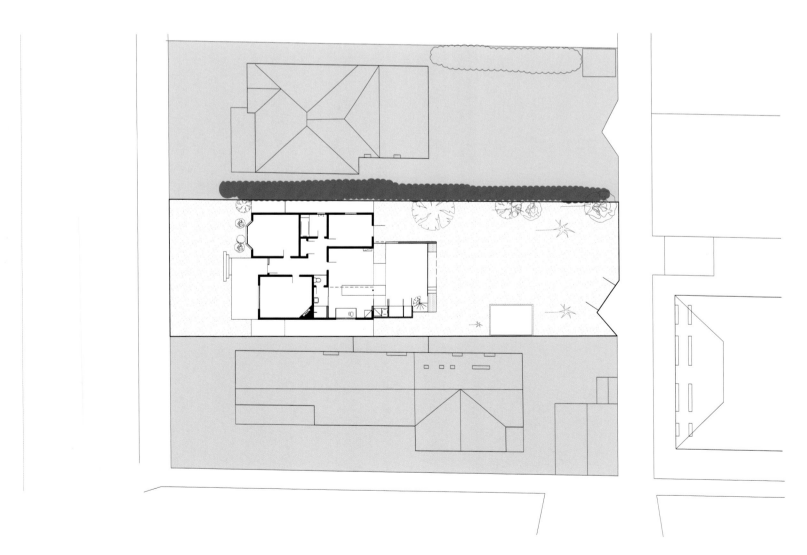

JD 住宅
JD House

Although the requirement was the two bedrooms and their respective bathrooms, the social area must be large enough with the possibility of being adapted to different uses as the owner frequently entertains many friends, which made the particularity of the project.

The program was resolved by crossing two prisms perpendicularly and placing it in a forest clearing between the trees. The strong slope of the land was exploited to hide part of the program, reducing the presence of the built volume.

　　该项目的特殊性在于，虽然要求设计的是两间卧室和各自的浴室，但由于主人常常招待许多朋友，因此要求社交空间必须足够大，以适应不同用途的可能性。

　　这个方案通过跨越互相垂直的两面棱镜和放置在树间的森林空地中来解决。土地牢固的边坡被用来隐藏设计的一部分，缩小了建筑的可见体积。

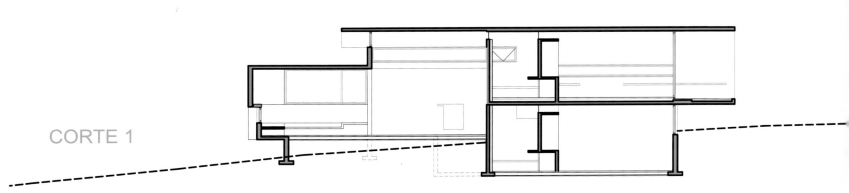

CORTE 1

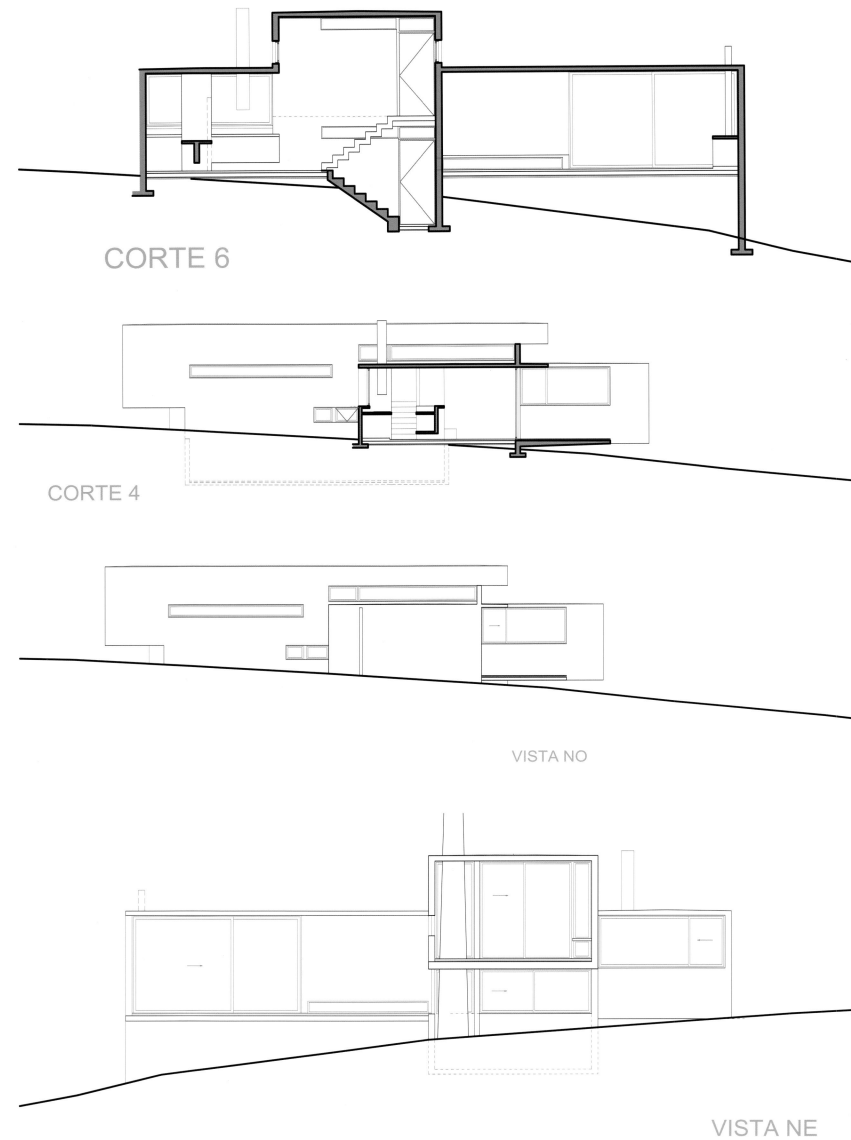

CORTE 6

CORTE 4

VISTA NO

VISTA NE

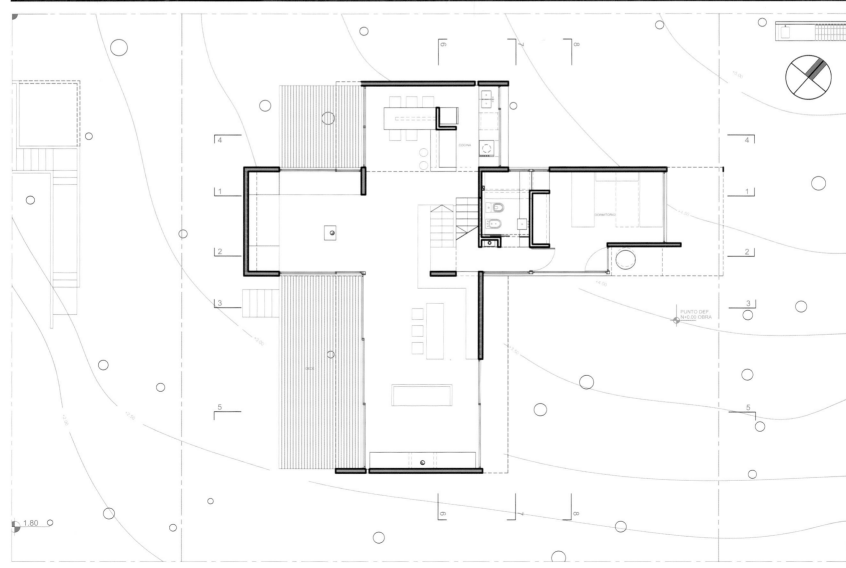

斯塔克霍夫改造住宅

Streckhof Reloaded

Urban Planning

The site is located in a settlement of detached single-family houses, characteristic for the 1970s. The traditional arrangement of functions for each storey: basement/ garage, ground floor / living space, attic floor/ bedrooms, has been transformed into a linear order. The so called "Streckhof", the original farm model for this area, has been adapted by stringing together different functions in one floor.

Function

Different functions are summarized in three structures similar to container. The main entrance from the street is marked by a canopy which leads to the fully glazed passageway and ends in an overhang at the garden side. Garages and storage space is located in the first of three containers. The connecting passage to the second container acts as porch.

The living space, gets natural light from three sides and is situated in the second container. A vegetable garden is connected directly to the kitchen.

The third container implies the intimate areas, far off the street and the entrance area. Two east orientated children's bedrooms share one bathroom lying in between. There's the parents bedroom orientated to the west. The terrace with the elongated pool is framed by the last two buildings and thereby wind and privacy protected.

城市规划

　　该项目位于具有 20 世纪 70 年代特色的独户住宅居住小区内。每层功能按照传统布局：地下室是车库、一层是生活空间、顶层是卧室，已被转化为一种线性的顺序。所谓的"Streckhof"，是当地原始的农场模型，已被改造成将不同的功能结区在地面上直线式分布。

功能

　　不同的功能区被安排在三个类似于集装箱的房子里。从街道上看，道路主入口处有一顶天蓬，一直通向住宅的玻璃走廊，最后延伸到花园边一个悬空的房间下面。第一个"集装箱"是住宅的车库和仓储室，连接第二个"集装箱"的通道则是住宅的走廊。

　　第二个"集装箱"是起居室，自然光线从三面射进来，菜园紧挨着住宅的厨房。

　　第三个"集装箱"离街道和入口区都很远。两个儿童卧室朝向东边，中间有共用的浴室，大人卧室朝向西边。带有方形水池的阳台位于后两个"集装箱"（起居室和卧室）之间，因此它的周围都被环绕着，隐私性很好。

cess is possible from the kitchen, the living room, the hallway as well as the parents' droom. The walkway consists of open and closed sequences that allow constant views o the half-open courtyards and each container. By planting both sides of the glazed ssageway with vine, the client, who is a hobby wine grower, gets his own piece of eyard into his house.

onstruction

make sure that the client can build as much as he is able to on his own, the outside lls are planned as brick construction. Even the insulation and the facade made of

acryl glass have been installed by the client's family. The roof construction consists of prefabricated timber elements with a foil sealing. The connecting passageway, also made of timber, acts as a bridge. Its facade consists of structural glazing without any mechanical fixing.

Facility Engineering

All living rooms are equipped with underfloor heating. The water is heated through a geothermal heating collector and two heat pumps. The fresh air is supplied via controlled ventilation with heat recovery. The swimming pool is heated by a solar collector on the rooftop.

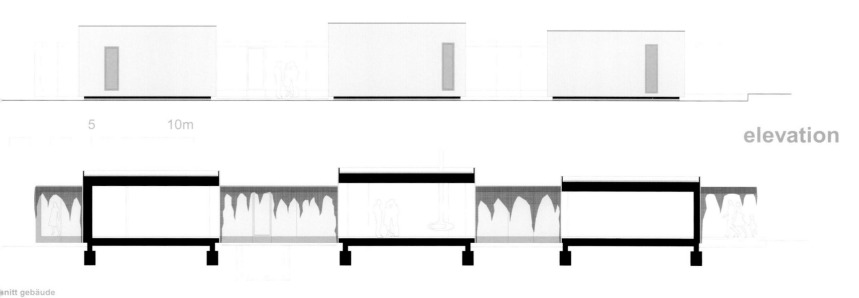

5   10m

elevation

nitt gebäude

5   10m

section

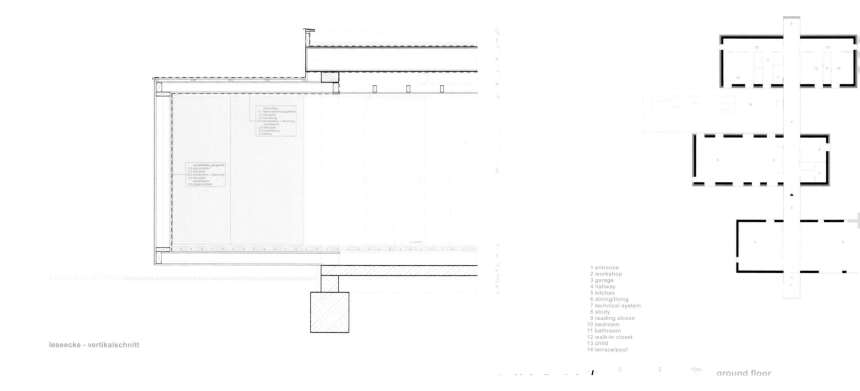

leseecke - vertikalschnitt

1 entrance
2 workshop
3 garage
4 hallway
5 kitchen
6 dining/living
7 technical system
8 study
9 reading alcove
10 bedroom
11 bathroom
12 walk-in closet
13 child
14 terrace/pool

0   5   10m   ground floor

从厨房、客厅、走廊以及大人卧室都可以进入到这里。通道由开放和封闭的顺序组成,这样就能看到半开放的庭院和每个"集装箱"。房屋主人是一位葡萄酒爱好者,在玻璃通道两边都种上了葡萄,这使房主在自己的住宅里拥有一片葡萄园。

构造

为了让用户尽可能多地亲自设计自己的房屋,设计师只用砖块初步地砌成了外墙。甚至连房屋的保温墙和由亚克力玻璃做成的立面,也是由用户自己安装的。屋顶由预先定做的、带有金属箔片封口的木质材料做成。作为桥梁的连接通道也是用木材做成的。它的立面由结构型玻璃组成,因此不需要任何机械进行固定。

设施工程

所有的房间都配备了地热供暖。水通过集热器和两个热泵加热。新鲜空气通过带有热回收功能的通风装置进入室内。游泳池则通过屋顶的太阳能集热器加热。

Franz 小屋

Franz House

This work is a summer house in the forest of Mar Azul. It was a small house of concrete of about 80 m² to be used mainly in summer, with only two bedrooms of minimum dimensions and a bathroom.

Take Advantage from What the Environment Offers

To recognize this particular microclimate was determinant to make choices in the aesthetic and constructive system of the house. It was necessary to gain light allows us to consider the entirely work as a "semi covered", resolving it by using big glass which provides: from the inside, great views to the sea and to the landscape; and, from the outside, reflects the nature allowing the work to blend gracefully with it.

这是是一个位于 Mar Azul 森林的避暑小屋，是一个约 80 m² 的混凝土小房子，主要在夏天使用，只有两个最小尺寸的卧室和一个浴室。

利用自然环境资源

认识到这一特殊的小气候是选择房子美观和结构系统的决定因素。有必要知道让我们将整个项目看作一个"半覆盖"型设计的观点，这可通过使用大型玻璃来完成：从里往外看，可以看到广阔的海景和风景；从外面看，反映大自然让此项目优雅地与之融为一体。

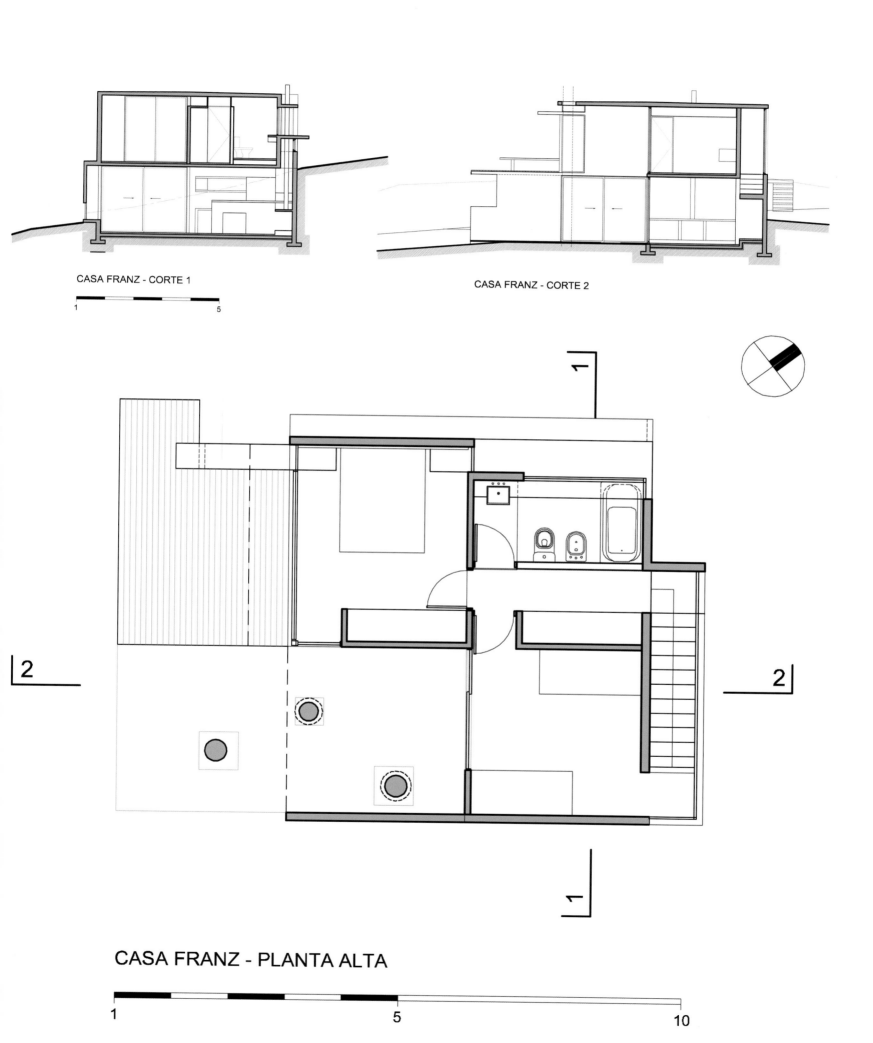

CASA FRANZ - CORTE 1

CASA FRANZ - CORTE 2

CASA FRANZ - PLANTA ALTA

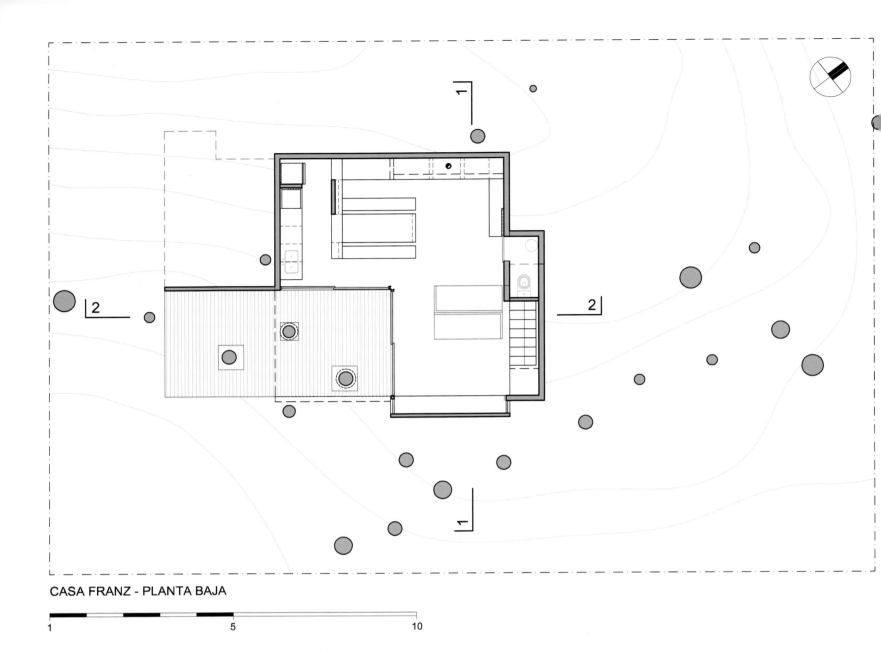

CASA FRANZ - PLANTA BAJA

1   5   10

Judicious Use of Available Resources

The expressiveness qualities of the exposed concrete and its properties of resistance and impermeability, make it unnecessary to use a surface finish, gaining a low budget in its execution without a future maintenance. On the other hand, the color and texture of the concrete provides a strong and mimetic presence at the same time, allowing the work to be in harmony with the landscape. In brief, an outer skin made by two materials: glass and concrete, resolve the integration with the landscape and gives us an answer to the functional, formal and structural issues of the surface finish and maintenance.

The Construction

The house is constructed of three basic materials: exposed concrete, glass and wood in the outdoor decks. The slabs of the different volumes are supported by walls and exposed concrete reversed beams, and are finished with a minimum slope in order to produce a faster rainwater runoff. It was used the same concrete of the other works in Mar Azul (H21 with fluidizer) a mixture with low amount of water that when forge create a more compact concrete. The floors are made of concrete screed cloths divided by plates of aluminum. The openings are of dark bronze anodized aluminum. The heating system, since there is no natural gas in the area, is solved by combining salamander, bottled gas stoves and electric stoves.

The Furniture

Except the beds, couches and chairs, the rest of the equipment of this house is solved in concrete.

明智利用现有资源

裸露的混凝土表现出的特质及其隔热和抗渗性能，让该项目无需外观抛光处理，以后也不需要维护，因此降低了预算成本。另一方面，混凝土的颜色和纹理，提供了一个强大的虚有外表，使该项目与景观协调一致。总之，由玻璃和混凝土两种材料构成的外墙解决了与景观的融合问题，对于外观抛光和维护的功能、形式和结构问题都交出了答卷。

结构

房子是由三种基本材料建成：裸露混凝土、户外阳台的玻璃和木材。不同大小的厚板由墙壁和裸露混凝土反梁支撑，并由一个最小坡度来完成，以产生更快的雨水径流。这也是在 Mar Azul 的其他项目一样的混凝土（有 fluidizer 的 H21）和少量水混合后形成更紧凑的混凝土结构。地板是混凝土整平布构成，由铝片分开，开口处是深色青铜铝阳极氧化的。由于该地区没有天然气，暖气系统是通过壁炉、瓶装气炉和电炉等解决的。

家具

除了床、沙发和椅子外，这座房子的其余部分的设备都是混凝土的。

Terrace

Kitchen

Dining

Bathroom

Living

Bedroom

Terrace

921 住宅
Villa 921

Iriomote-Island is a part of Okinawa in the south of Japan. With an area of over 90% covered by subtropical virgin forest, the entire island is designated as a national park. While sitting snug against a hill, at the front of the house a luxurious vision opens up overlooking the farmlands that spread around the site.

The necessary rooms included the living and dining room, a parent's bedroom, kitchen, toilet, a bathroom and a storage room that would be large enough for diving tools. The bedroom has two doors so that a separate children's room can be created. Leaving out the rain gutter, the gabled roof lets the rain water run over the facade to wash away

accumulating salt crystals.

The west side of the house opens up to the beautiful landscape. Wide eaves protect against the sun, which is five times stronger than on the main island. The spacious terrace is also used to wash and dry diving tools. During a typhoon, wind protection nets can be installed on the eaves. The usable area amounts to about 70 m². This is by no means large, but thanks to the amazing views of the landscape, there is never a feeling of narrowness. Embedded in the surrounding area, it is a simple life in a little house. The new house has already become a part of Iriomote-Island.

　　西表岛是日本南部冲绳的一部分，超过 90% 的面积为亚热带原始森林，整个岛被指定为国家公园。房子坐落在舒适的山边，前面是一片奢华的景色，俯瞰着周围一望无际的农场。

　　必备的房包括客厅、饭厅、家长卧室、厨房、厕所、浴室和用于放置潜水用具的大贮藏室。卧室有两个门，以便可以隔成一个单独的儿童房。没有了檐槽，山形的屋顶让雨水在外层上流动并洗去积盐晶体。

　　房子的西边向美丽的风景敞开。宽屋檐遮挡了太阳光，这里的太阳光强度是主岛的五倍。宽敞的阳台也可用于清洗及晾干潜水用具。刮台风时，可以在屋檐上安装防风网。可用面积达 70 m²。这里不大，但由于风景美丽，因此决不会感觉狭小。融入周边地区中，这是一个小房子里的简单生活。这座房子已经成为西表岛的一部分。

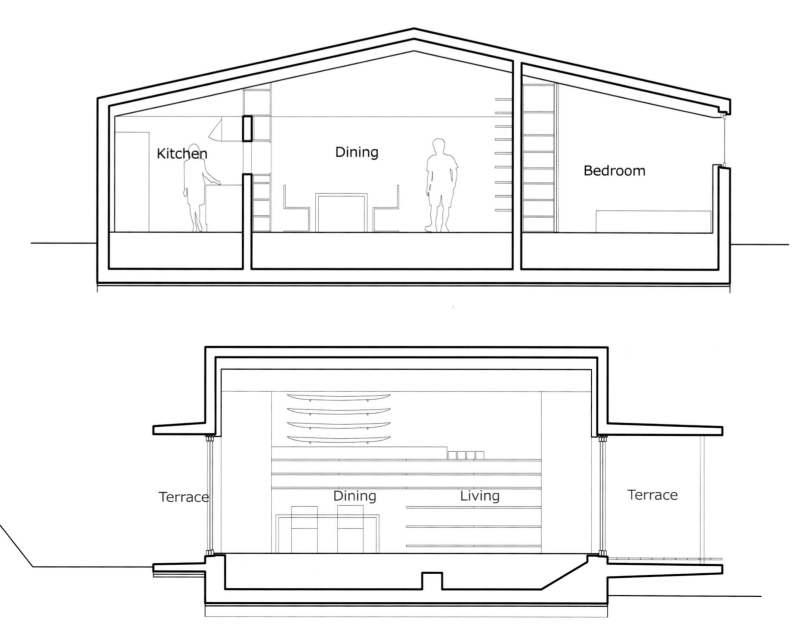

Kitchen

Dining

Bedroom

Terrace

Dining

Living

Terrace

卢加诺湖别墅

Lake Lugano House

Lying on the slope of a hill, on the shores of Lake Lugano, the villa consists of two volumes organized on different levels due to the particular topography of the site.

A polygonal shaped glass pavilion with rounded edges stands above a linear underground block. The living and dining room, the kitchen and storage spaces are located in the pavilion, while bedrooms, bathrooms and garage are in the lower level. Each level relates itself with independent outdoor spaces, which are closely related with the interiors.

The glass pavilion overlooks two much defined areas: the first, toward the mountain, is a very private zone resulted in the area between the property line and the building setback line according to the local building code. The second is a garden overlooking the lake. In the same way, the bedrooms face a garden enclosed by the building and the perimeter wall.

The ring, obtained between the perimeter wall above and the pavilion, amplifies the interior space, with seems much larger than what it actually is. The ring-like space, that embraces the building on the north side, grants constant ventilation and natural light to the living areas, also due to the white cladding of the perimeter wall and white gravel which reflect the sunlight coming from the south, a night-time artificial light scene is the ideal reverse field for the lake panorama.

位于卢加诺湖畔的山坡上，由于地形的特殊性，该别墅由两栋不同标高的体量组成。

一个圆角的多边形玻璃楼阁建在线性地块上。客厅、饭厅、厨房和储物间都在这里，而卧室、浴室和车库则在较低楼层，每层本身都有独立的、和室内紧密相连的户外空间。

玻璃楼阁俯瞰两大特定区域：第一，面向大山，是一个非常私人的区域，是根据当地建筑规范规定的建筑红线和建筑后退红线之间的场地。第二是一个俯瞰卢加诺湖的花园。以同样的方式，卧室面对着由建筑和围墙围起来的花园。

上述围墙和楼阁之间形成的环形，放大了室内空间，让其看上去似乎比实际尺寸大很多。环绕北边建筑的环形空间，保障了生活区的持续通风和自然光线。由于围墙的白色覆面和反射南边阳光的白色碎石，一个夜间人造灯光场景是湖水全景理想的反向场。

All the additional functions of the pavilion are contained in a central lacquered wood block, which acts as a sort of a thick penetrable wall that separates the kitchen from the living room without dividing the space with doors, and in which are located the powder room, the kitchen, the stairs, bookcases, all mechanical systems and the technological and audio-video equipment.

Great attention is given to the environmental aspects, such as the use of geothermal energy, roof gardens, the rain-water collection system, the choice of highly efficient low-emittance glass insulated with argon gas, to optimize the thermal efficiency of the shell and the use of natural sun shading as the placement of deciduous trees in the south-west area of the building.

楼阁的所有附加功能都包含在一块中央涂漆的木块中，该木块充当可穿透的厚墙，不需要门就可将厨房和起居室隔开，并且里面还有盥洗室、厨房、楼梯、书柜、所有机械系统和技术及音视频设备。

该项目高度重视环境问题，如地热能的使用、屋顶花园、雨水收集系统、高效低辐射氩气隔离玻璃等，由于在建筑西南部种植了落叶乔木，优化了楼阁外壳的热效率和自然遮阳的使用。

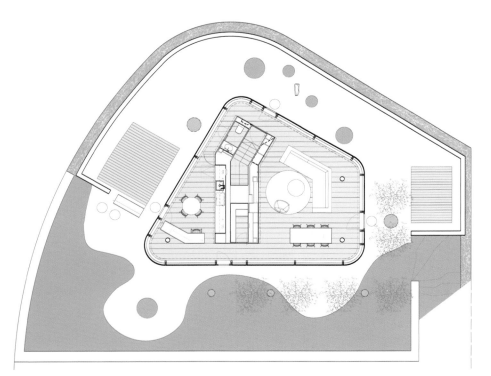

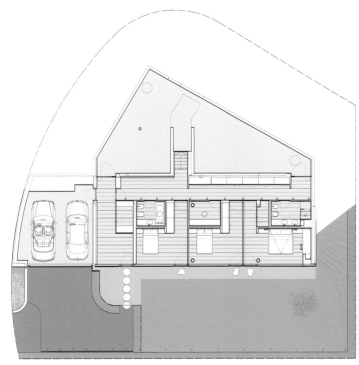

梅里涅米夏日住宅

Summerhouse Meriniemi

This little cottage is one of a group of buildings used by three generations of the same family. It is set in a rocky hollow beside the open sea, right on the edge of the forest. The group of buildings is linked together by a shared terrace and a turf-roofed playhouse in the middle of the central courtyard.

The end of the new building, which opens towards the sea, is built entirely of glass. The interior is protected from direct sunshine by a translucent roof-covering which cuts off the sunlight. Each of the long walls has a narrow strip window at the top to let morning and evening sunshine into the lofty living space.

The wedge-shaped section of the small building gives enough space to sleep a family of five in the cabin-like bedroom with its high-level balcony. In relation to the amount of interior space there is a large quantity of covered exterior space for summer holidays. The terrace, with its sheltering pine-batten louvers, is used for untangling fishing nets and for preparing food, always most enjoyable done out of doors.

The building and the terrace are supported on longitudinal beams of laminated veneer lumber (Kerto LVL). The foundations are constructed on the rocky ground as few concrete columns as possible. The main members carrying the roof structure are also in LVL and the heat insulation is of wood-fibre.

这座小屋是三代同堂的一组住房之一，位于海边的一个岩石山谷中，就在森林的边缘。这组住房通过由公用阳台和中央庭院中间的草皮屋顶的儿童游戏屋连在一起。

朝向大海的新建筑的一端全是用玻璃建造的，内部通过一个半透明遮阳的屋面避免阳光直射室内。每面长墙壁顶部有一个狭长的窗口，早晨和傍晚时可让阳光进入高耸的生活空间。

小屋的楔形部分，像客舱一样的卧室以及高空阳台，有足够的空间容纳一个五口之家。相对室内空间而言，有大量覆顶的外部空间供夏天休假使用。带有起遮蔽作用的松树板条百叶窗的阳台，用于清理渔网和准备食物，是户外最愉快的活动。

建筑和阳台由叠层薄木（Kerto LVL）的纵向梁支撑，地基是建在岩石地上，尽可能少用混凝土柱。支撑屋顶结构的主材料依然是LVL，隔热材料是木材纤维。

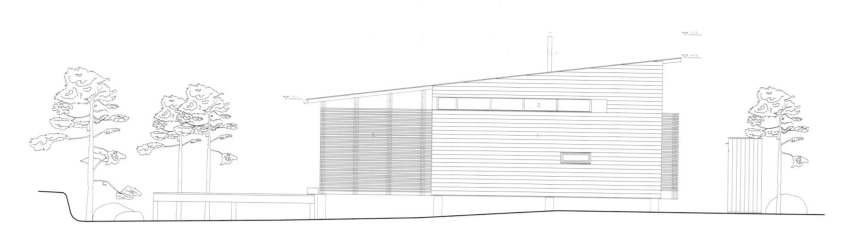

JULKISIVU ITÄÄN

JULKISIVUMATERIAALIT:

1. Ruskea puunsuoja
2. Lasi
3. Puusäleikkö (mänty)
4. Musta kattohuopa

0m 5m

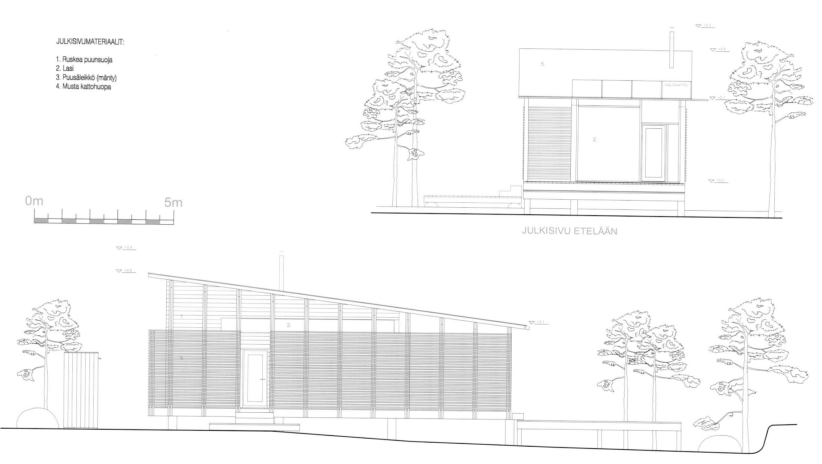

JULKISIVU ETELÄÄN

JULKISIVU LÄNTEEN

JULKISIVUMATERIAALIT:

1. Ruskea puunsuoja
2. Lasi
3. Puusäleikkö (mänty)
4. Musta kattohuopa

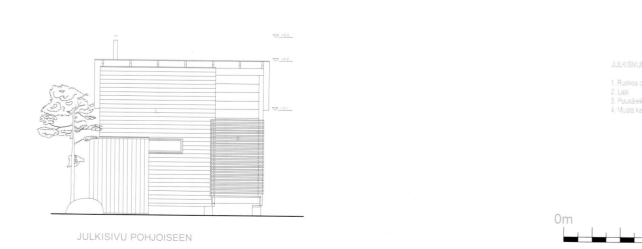

JULKISIVU POHJOISEEN

0m 5m

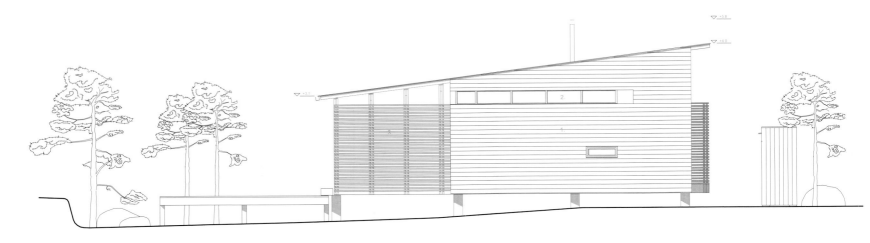

JULKISIVU ITÄÄN

JULKISIVJMATERIAALIT:

1. Ruskea puunsuoja
2. Lasi
3. Puusäleikko (mänty)
4. Mustakatto huopa

0m   5m

VALOKATTO

JULKISIVU ETELÄÄN

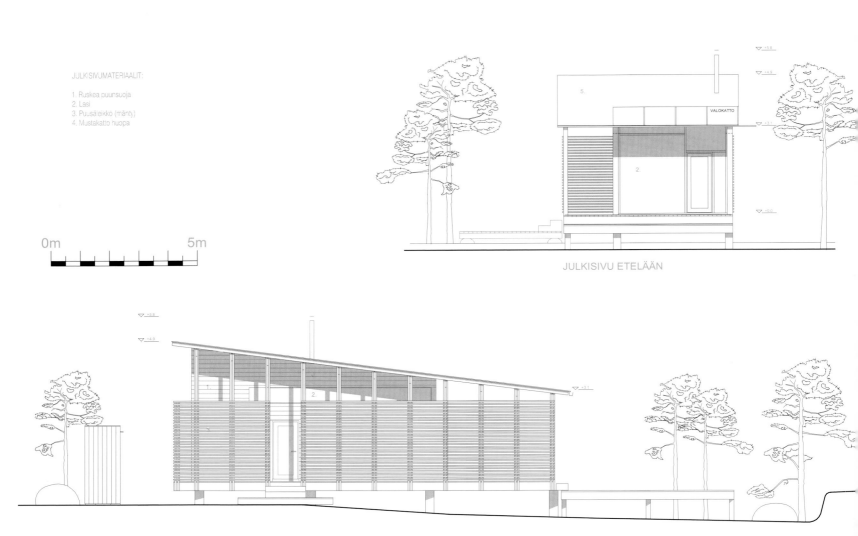

JULKISIVU LÄNTEEN

JULKISIVU POHJOISEEN

JULKISIVUMATERIAALIT

1. Ruskea puunsuoja
2. Lasi
3. Puusäleikkö (mänty)
4. Musta kattohuopa

0m   5m

茵尼斯基薄荷房屋

Infiniski Menta House

The house of 100 m² is designed in the shape of a cube organized on one level, including a central living area with an open kitchen, two rooms and a common bathroom.

Located in the country side of the Tarragona Province, the project was thought of as a weekend house, which needed to be easy to use, efficient and which would take full advantage of its natural surroundings. The house was designed as a "living box" which can be "opened", "closed", "switched on", "heated", "cooled down" efficiently, easily and rapidly. The house is designed as a cube – rational and functional – where the transition between exterior and interior areas is as fluid as possible.

The facade system uses Corten steel panels which create an eye – catching contrast with its natural surroundings. The shutters play an important role both for the aesthetic value of the house and its energetic efficiency. The shutters, which are completely integrated within

the facade, use perforated panels of Corten steel. The panels are perforated with the shape of mint leaves which project luminous forms in the interior during day and in the exterior at night, a little like a "light box". Those perforated shutters, when closed also work as a solar protector allowing air and light to pass through.

Instead of using shipping containers, Infiniski decided to use four prefabricated (inhouse) metal modules. The walls are composed of an eco-friendly insulation layer made of sheep wool, and cellulose panels. With the joint action of its bioclimatic design, the use of eco-friendly building materials, the use of renewable energies (in this case biomass heating and solar panels), the house gains very high thermal efficiency and was recognized with the highest level of energetic efficiency (Certification A) by the Regional Institute of Energy ICAEN. This allows a reduction up to 60% of energetic consumption.

该房屋占地一层，共 100 m²，设计成立方体的形状，包含了中间的生活区和一个开放性厨房，两间房间以及一个公共盥洗室。

该项目坐落于塔拉戈纳省的乡间，被当成周末小住之处，因而必须易于使用，且高效，并充分、有效地利用周围的自然环境。房子被设计成为一个"活箱"，可以简单、快速、有效地"打开"、"封闭"、"通电"、"加热"或"冷却"。并且其立方体设计无论是在合理性上还是在功能性上都保证了室内外尽可能流畅地过渡。

它的外立面系统使用了柯尔顿钢质面板，与自然周围环境形成吸引眼球的鲜

明对比。百叶窗则在美观和节能方面都发挥了重要作用，它们完全融入了外立面，也使用了柯尔顿钢质穿孔面板。这些面板上有薄荷叶形状的洞孔，白天为室内带来光亮，晚上则将光线投射到室外，有点儿像一个 "光箱"。这些穿孔百叶窗关闭时也可作为一个太阳能保护器，允许空气和光线通过。

茵尼斯基没有使用船运集装箱，而是使用了四个预制（自身）金属模块。墙壁由环保的羊毛隔热层和纤维板构成。在生物气候设计、环保建筑材料和可再生能源使用（本案中体现的是生物质能供热和太阳能电池板）的共同作用下，这幢房屋获得非常高的热效能，得到 ICAEN 区域能源研究所最高级别的能量效率（A 认证）的认可，可以减少高达 60% 的能量消耗。

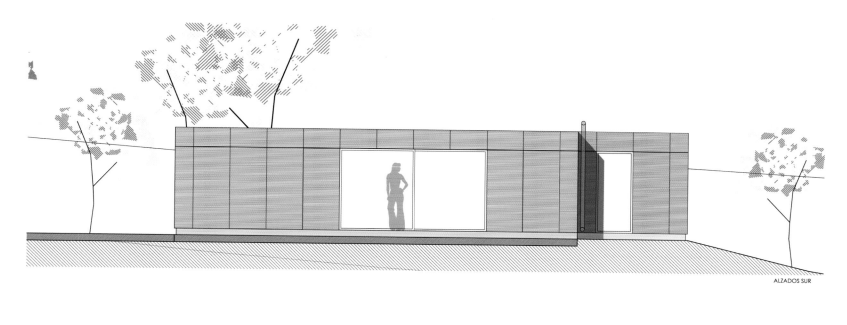

ALZADOS SUR

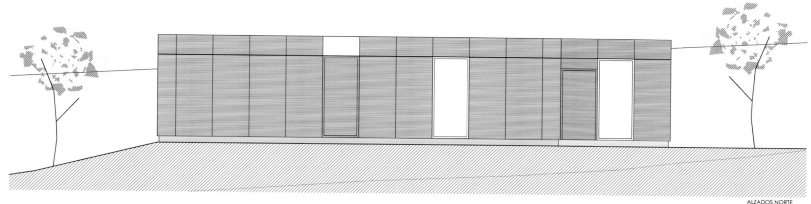

ALZADOS NORTE

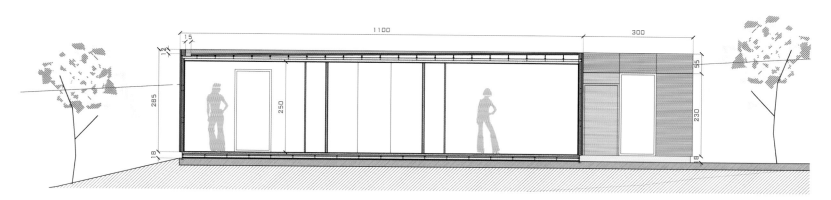

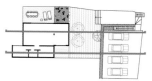

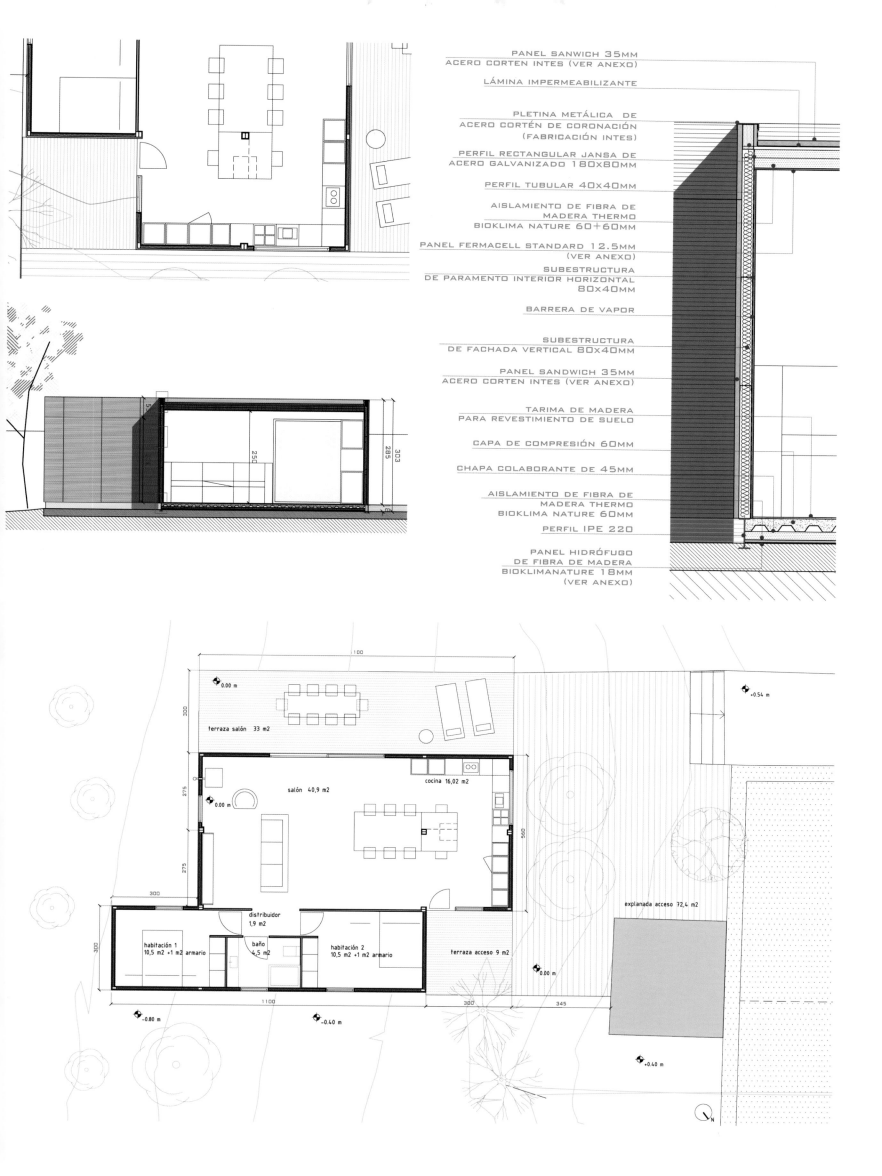

PANEL SANWICH 35MM
ACERO CORTEN INTES (VER ANEXO)

LÁMINA IMPERMEABILIZANTE

PLETINA METÁLICA DE
ACERO CORTÉN DE CORONACIÓN
(FABRICACIÓN INTES)

PERFIL RECTANGULAR JANSA DE
ACERO GALVANIZADO 180x80MM

PERFIL TUBULAR 40x40MM

AISLAMIENTO DE FIBRA DE
MADERA THERMO
BIOKLIMA NATURE 60+60MM

PANEL FERMACELL STANDARD 12.5MM)
(VER ANEXO)

SUBESTRUCTURA
DE PARAMENTO INTERIOR HORIZONTAL
80x40MM

BARRERA DE VAPOR

SUBESTRUCTURA
DE FACHADA VERTICAL 80x40MM

PANEL SANDWICH 35MM
ACERO CORTEN INTES (VER ANEXO)

TARIMA DE MADERA
PARA REVESTIMIENTO DE SUELO

CAPA DE COMPRESIÓN 60MM

CHAPA COLABORANTE DE 45MM

AISLAMIENTO DE FIBRA DE
MADERA THERMO
BIOKLIMA NATURE 60MM

PERFIL IPE 220

PANEL HIDRÓFUGO
DE FIBRA DE MADERA
BIOKLIMANATURE 18MM
(VER ANEXO)

terraza salón 33 m2

+0.54 m

0.00 m

salón 40,9 m2

cocina 16,02 m2

distribuidor
1,9 m2

baño
4,5 m2

habitación 1
10,5 m2 +1 m2 armario

habitación 2
10,5 m2 +1 m2 armario

explanada acceso 72,4 m2

terraza acceso 9 m2

0.00 m

-0.80 m

-0.40 m

+0.40 m

N

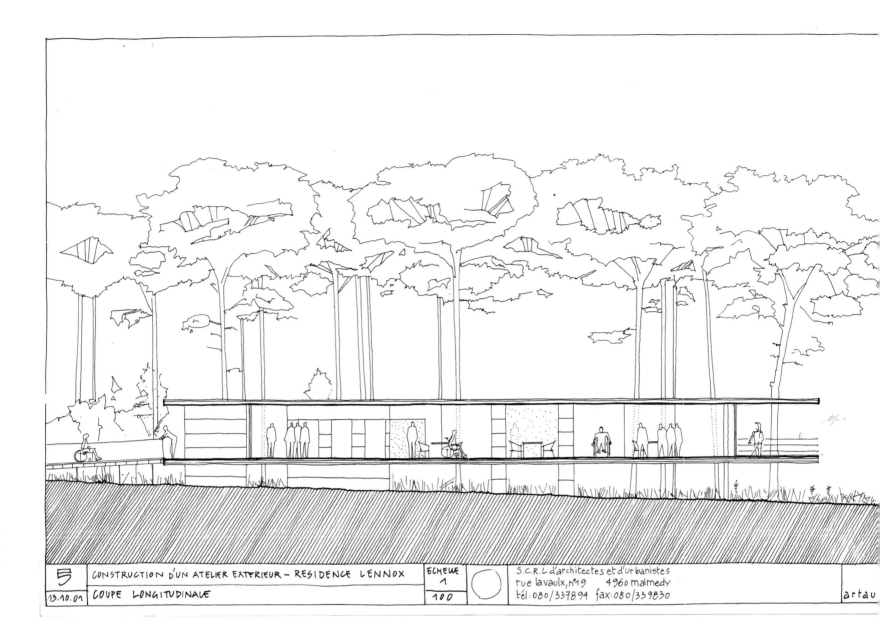

CONSTRUCTION D'UN ATELIER EXTERIEUR - RESIDENCE LENNOX | ECHELLE 1 | ○ | S.C.R.L d'architectes et d'urbanistes rue lavaulx, n°19   4960 malmedy tél:080/337894 fax:080/339830 | artau
13.10.01 | COUPE LONGITUDINALE | 100 | | |

伦诺克斯住宅

Résidence Lennox

The building accommodates epileptics with poor mobility. It is located within a pinewood at the edge of a reserve forest. The room with windows all around offers a permanent relation with the surrounding nature and gives them the impression of being outdoors. The "vessel" made of glass with a structure of wood and iron is based on pillars and floats therefore above the nature. A wooden footbridge meanders under the pine trees and assures the link to the existing building.

The whole on concrete piles / coffered flooring and beam roofing of the brand Kerto and ground beams / pillars and support of cantilever made of steel / horizontal finishing with panels and tiles / vertical finishing outside made of glass / vertical finishing inside out of linoleum and plasterboard.

Passive solar system, double glazing k1,1/ solar protection=existing pine vegetation / insulation of flooring and roofing made of cellulose flock / air-tightness testing.

该住宅供行动能力较差的癫痫患者居住，它坐落于一个森林保护区的松林边缘。四周带有窗户的房间提供了与周围的大自然长期不变的关系，为患者带来身处室外的感觉。由玻璃制成的具有木头和铁结构的"器皿"放置于柱石和漂浮物之上，因而高于地面。木制的人行桥在松树下蜿蜒，确保了与现有建筑的联系。

整个建筑物由混凝土桩，方格地板和柯特牌屋面梁板，地面梁柱柱子，钢质

悬臂梁，经过抛光处理的横向面板、瓷砖，外面的垂直玻璃抛光面，用油毡和石膏板修饰的室内构成。

现有的松树植被、绝缘的地板、纤维板制成的屋顶和气密性测试等构成了该住宅的诱导式太阳能系统以及双层玻璃 k1,1 带来的太阳能保护。

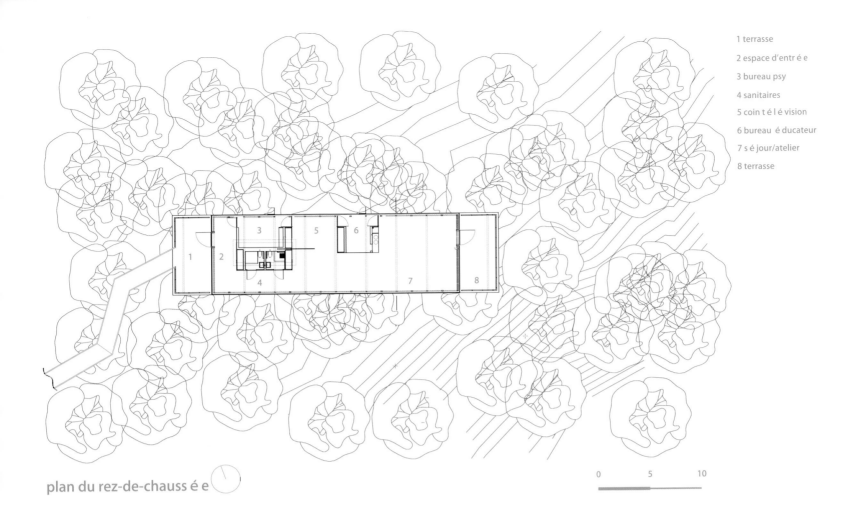

1 terrasse

2 espace d'entrée

3 bureau psy

4 sanitaires

5 coin télévision

6 bureau éducateur

7 séjour/atelier

8 terrasse

plan du rez-de-chaussée

0   5   10

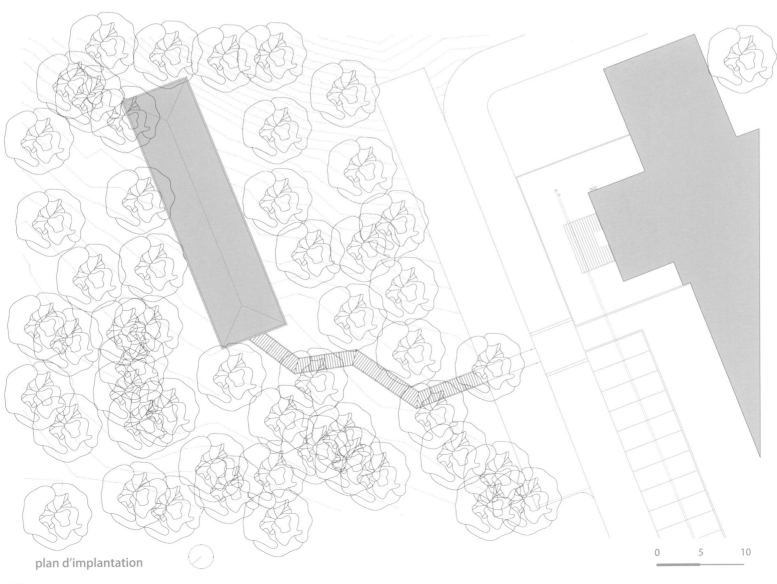

plan d'implantation

0   5   10

树屋

The Treehouse

The major image of the design is a sheet of paper that is pleated and encloses both interior and exterior spaces. The Treehouse is based on five elements: two cabins on different levels, connecting terraces, a staircase and a connecting roof.

The Cabins

In the lower cabin is a coffee lounge, pantry, restroom and technology room. The upper room is dedicated to meetings and other events that are appropriate to this exquisite space among the trees.

Benches covered with soft pillows surround the space resulting in a comfortable and laid-back feeling, encouraging one to stay longer.

Construction

The cabins and upper terraces rest on 19 angled steel stilts as the lower landing of the staircase hangs on steel cables that are connected to a pine tree. Each steel stilt is connected to the ground below with foundation screws that have minimal impact on the forest floor.

　　该设计的主要形象是一张打褶且包围内外空间的纸。树屋包括五部分：位于不同层上的两个木屋、相连的阳台、楼梯和相连的屋顶。

木屋

　　低层木屋设有一个咖啡厅、食品储藏室、洗手间和科技室。高层房间是专为会议和适合这个位于树间的精美空间的其它活动而准备的。

　　空间四周的长凳上覆盖着柔软的枕头，营造了舒适悠闲的休息空间，鼓励人们多停留一下。

结构

　　几根倾斜 19° 的钢柱支撑着小木屋和高层阳台，楼梯的低层平台挂在几根钢丝绳上，钢丝均连接到一棵松树上。每根钢柱连接到地面上，并以地基的螺丝钉固定，把对森林地面的影响降到最小。

Heating and Lighting

All equipments use the existing supplies (green electricity and water purification) of the nearby museum or are self-sufficiently (heat pump).

The newest techniques are used for the heating and cooling system. A heat pump unit which is situated in an underground space extracts heat from the air. The warmth is transported through tubes which are led through one of the steel columns into the cabins. After converting the warmth to a fluid, tubes will transport it to the heating-units under the benches. The result of this low temperature heating system is a nice, comfortable climate and the energy cost remains very low. In summertime the system can be used for cooling as well. This system is CO_2 neutral.

The balanced ventilation system is totally demand-based with supply and abstract units integrated in the benches. The ventilation and heating systems are continuously monitored. To minimize the energy consumption, all lights are LED. In the toilet, daylight and movement-sensors are integrated in the LED armatures.

供暖和照明

所有设备使用附近博物馆的现成供应（环保电力和水净化）或者自给自足（热泵）。

最新科技应用到加热和冷却系统中。热泵机组位于地下空间，从空气中提取热量。暖气是通过穿过其中一个钢柱的管道传输到小屋。暖气转换成流体后，管道将其运输到长凳下的加热组件中。这种低温暖气系统的效果是极好的、舒适的气候，并且能源成本能保持极低水平。在夏季，该系统也可用于冷却，它是二氧化碳平衡的。

平衡通风系统完全基于需求供应，是融入到长凳中的抽象单元。通风和加热系统都是不间断地被监控中的。为了减少能耗，所有的灯都是 LED 灯。在厕所里，太阳光和运动传感器集成在 LED 电枢上。

Centre Léonce Georges

Chauffailles, a village in the French Bourgogne, needs a new multifunctional community centre that will be named Léonce Georges, in memoriam of a local hero.

The project recuperates some of these rural constructions, like granaries, barns, warehouses – rather large shelters mainly used for storing grain or machinery normally with a single opening facing the yard. These are anonymous and simple architectures, completely utilitarian, built with a specific shape and size for better storage, the same way the materials and sources chosen are local and accessible. This is where their beauty is held: in their pure utility, in their radical symbiotic and productive relationship which are primitive forms of sustainability.

The original L-shape construction is kept and full rehabilitated, keeping the typical elements such us the big wooden beams, the stone walls and the tiled roof. The addition, located in the adjoined yard, completes this volume with a new volume made of the local wood (Douglas). One big door opens outwards to the fields.

The new volume is built in the way the old wooden barns in La Bourgogne were built: the different sides and faces are supported by a braced structure which becomes sheathed. It is a dry construction, half prefabricated and dismountable, willing to recuperate the social character of the rural constructions.

The new centre Léonce Georges in Chauffailles is a facility designed to respond the need of diaphanous spatiality to hold large community meetings as weddings, receptions, dancing parties, etc. Appropriate light and acoustics as well as comfortable climatic conditions are needed. The use of local materials blends with the landscape, the village and the people.

绍法耶村是法国勃艮第的一个村庄，这里需要一个新的多功能社区中心，为纪念当地的一位英雄，该中心将被命名为雷昂斯·乔治斯。

该项目复原了一些农村建设，如粮仓、牲口棚、仓库——大型的用于储存谷物或机器的棚子，通常只有一个面向院子的出入口。这些都是无特证、简单、完全实用的建筑。为了更好地存储物品，它们常以特定的形状和尺寸建成，也选取了当地的建筑材料和资源。这就是它们的美丽之处：纯实用性，基本的共生和生产关系，这种关系也是可持续发展的原始形态。

该建筑保持并修复了原来的 L 形结构，保留了其典型的建筑元素，如大木梁、石墙和石板瓦屋顶。另外，位于相邻的院子里，以当地木材（Douglas）建造而成的新建筑完善了整体建筑结构。有一扇大门朝外面的农田开着。

新建筑是以勃艮第旧木谷仓的建造方式建成：正面和各侧面由支架结构支撑，并由其覆盖住。这是一种干式施工的建筑，采取半预制方法并可拆卸，意在复原农村建设的社会特色。

绍法耶的新雷昂斯·乔治斯中心的设计旨在应对举办大型的社区聚会，如婚礼、欢迎会、舞会等，需要适当的灯光、音响以及舒适的气候条件。当地材料的使用使其融入周围的景观、村子和村民。

鸟之岛屋
Birds Island

We have applied an integrated strategy of developing a zero-energy house that seamlessly dovetails the economic and environmental advantages of environmentally friendly living. The environmental and economic features of this way of living do not conflict with our client's lifestyle, rather it furthers their ability to comfortably enjoy their time at home.

While providing an expansive outdoor living deck that spans the whole of the site, the primary living space is concentrated inside cooled zones. A maximum amount of economical and energy efficient floor area is created and sheltered from the elements by a dynamic tensile structure. The traditional relationship between indoor and outdoor has been shifted, allowing for comfort while free of the bonds of traditional walls.

Extended living spaces are arranged separately from each other and bisected by landscaped areas with local vegetation. The surrounding tensile fabric flows through the interior, shaping and imbuing the spaces with sublime shading and view patterns.

我们采用了一种开发零能源房屋的综合策略，这种房屋完美结合了经济和环境的环保生活优势。这种生活方式的环境和经济特色与我们客户的生活方式并不冲突，反而增强他们在家时所享受到的舒适度。

在提供一个跨越整个场地的广阔的户外甲板的同时，主要生活空间集中在冷却区。通过动态拉伸结构创造了最大化的经济和高效节能的建筑面积。室内外之间的传统关系已经发生改变，让人感到舒适的同时摆脱了传统墙体的束缚。

扩建的生活空间被当地植被和景观一分为二，彼此分开。周围的拉伸组织流过内部空间，以庄严的阴影和风景图案塑造空间。

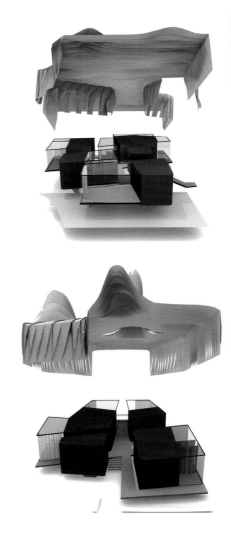

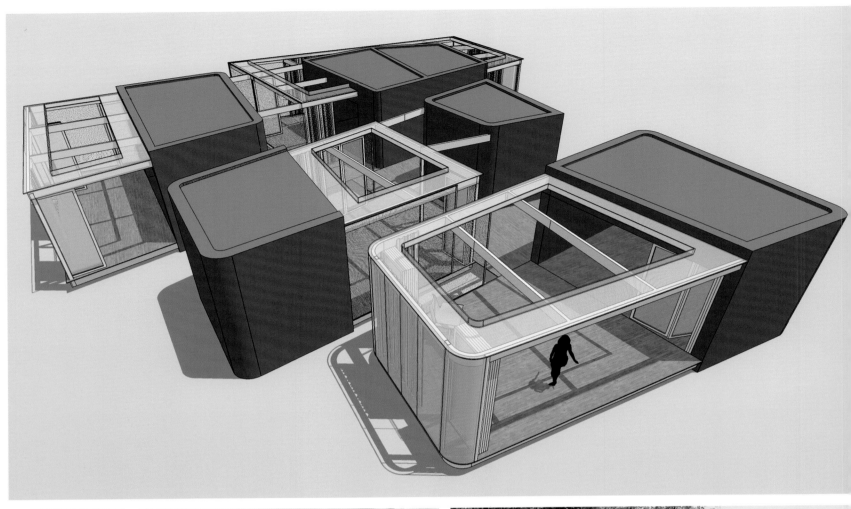

Architect · OFIS Architects   Client · Regie Immobiliere de la Ville de Paris   Area · 1981 m²   Photography · Tomaz Gregoric

巴黎篮子公寓

Basket Apartments in Paris

Urban Plan Conditions

The project is located on a long and very narrow site in Paris's 19th district. On the northeast, new Paris tram route is passing along the site. The site is bordering with tram garage on the southwest, above which is a football field. The first 3 floors of the housing will inevitably share the wall with the tram garage.

Program – Student Dormitory with 192 Studios

The major objective of the project was to provide students with a healthy environment for studying, learning and meeting. Along the length of the football field is an open corridor and gallery that overlooks the field and creates a view to the city and the Eiffel tower. This gallery is an access to the apartments providing students with a common place. All the studios are the same size and contain the same elements to optimize design and construction: an entrance, bathroom, wardrobe, kitchenette, working space and a bed. Each apartment has a balcony overlooking the street.

Design Concept

Narrow length of the plot with 10 floors gives the site a significant presence. Each volume contains two different faces according to the function and program.

The elevation towards the street contains studio balconies-baskets of different sizes made from HPL timber stripes. They are randomly oriented to diversify the views and rhythm of the facade. Shifted baskets create a dynamic surface while also breaking down the scale and proportion of the building.

城市规划条件

该项目座落在巴黎第 19 区一个狭长的场地。在东北面，新巴黎电车轨道沿着此区域而过。该处所西南面毗邻电车车库，上面是一个足球场。房屋的头三层将不可避免地与电车车库共用一面墙壁。

项目——有 192 间单间公寓的学生宿舍

该项目的主要目标是为学生提供一个良好的研究、学习和会议环境。在足球场边有一个开放的可俯瞰城市全景和埃菲尔铁塔的走廊，走廊通向公寓，为学生们提供了一个公用空间。所有单间公寓的大小相同，且均包含入口、卫生间、衣柜、厨房、工作空间和一张床，每个公寓都有一个可以俯瞰街道的阳台。

设计理念

10 层楼的狭长体量让此地成为一个显著的存在。每栋建筑根据功能和设计包含两个不同的外观：

面向街道的建筑立面设有单间公寓的阳台——由 HPL 木条制作的大小不同的"篮子"。它们的朝向是随机的，提供了多样化的景色和外观的节奏感。篮子的转换创建了一个动态的外观，同时也分解了建筑的规模和比例。

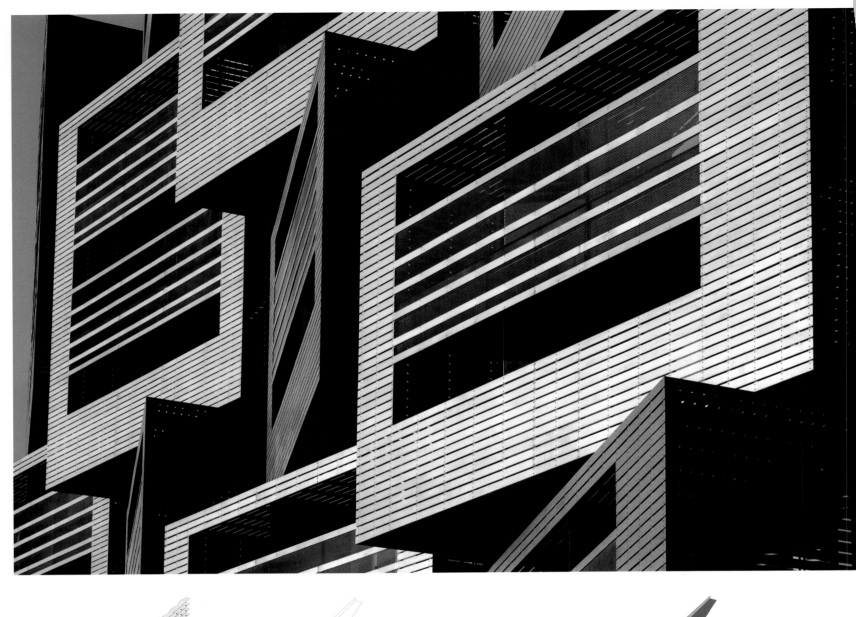

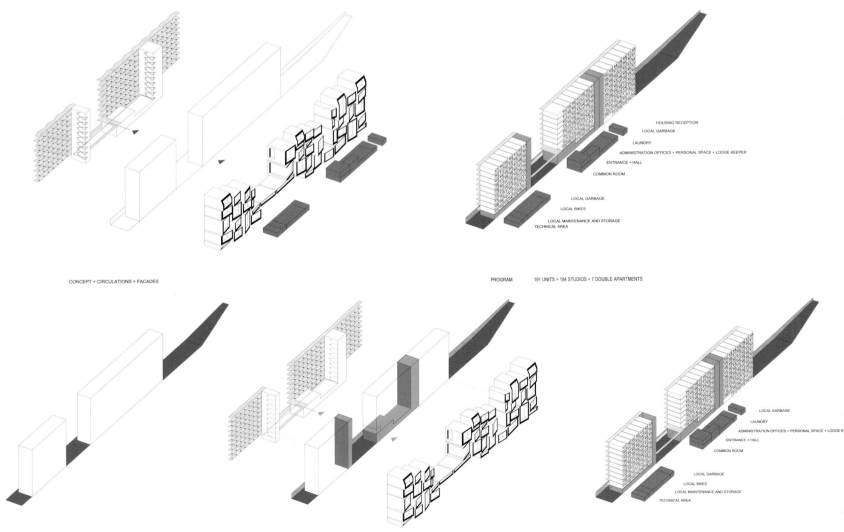

CONCEPT + CIRCULATIONS + FACADES

PROGRAM   191 UNITS > 184 STUDIOS + 7 DOUBLE APARTMENTS

HOUSING RECEPTION
LOCAL GARBAGE
LAUNDRY
ADMINISTRATION OFFICES + PERSONAL SPACE + LODGE KEEPER
ENTRANCE + HALL
COMMON ROOM

LOCAL GARBAGE

LOCAL BIKES

LOCAL MAINTENANCE AND STORAGE
TECHNICAL AREA

URBAN MASTER PLAN DEFINED VOLUME

CONCEPT + CIRCULATIONS + FACADES

PROGRAM   191 UNITS > 184 STUDIOS + 7 DOUBLE APARTMENTS

LOCAL GARBAGE
LAUNDRY
ADMINISTRATION OFFICES + PERSONAL SPACE + LODGE K
ENTRANCE + HALL
COMMON ROOM

LOCAL GARBAGE

LOCAL BIKES

LOCAL MAINTENANCE AND STORAGE
TECHNICAL AREA

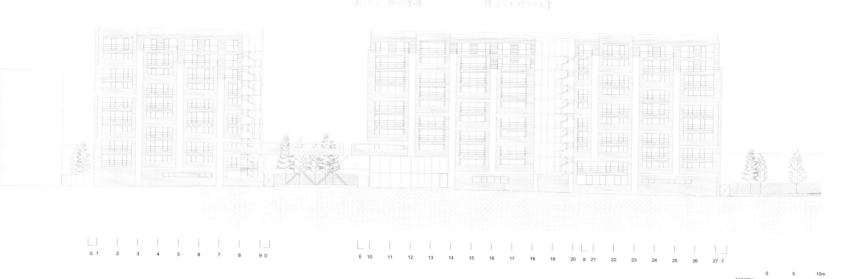

G 1 2 3 4 5 6 7 8 9 D   E 10 11 12 13 14 15 16 17 18 19 20 B 21 22 23 24 25 26 27 F

FACADE H   0   5   10m

The elevation towards the football field has an open passage walkway with studio entrances enclosed with a 3D metal mesh. Both volumes are connected on the first floor with a narrow bridge which is also an open common space for students.

Sustainable Efficiency

The building is energy efficient to accommodate the desires of Paris' sustainable development efforts. The objectives of energy performance and the construction timetable were met by focusing on a simple, well insulated and ventilated object that functions at its best year round. Accommodations are cross ventilating and allow abundant day lighting throughout the apartment. External corridors and glass staircases also promote natural lighting in the common circulation, affording energy while also creating comfortable and well lit social spaces. The building is insulated from the outside with an insulation thickness of 20cm. Thermal bridge breakers are used on corridor floors and balconies to avoid thermal bridges. Ventilation is controlled by double flow mechanical ventilation, providing clean air in every apartment with an optimum temperature throughout the year. The incoming air also reuses heat from the exhaust air. The roof is covered with 300 m² of photovoltaic panels to generate electricity. Rainwater is harvested on site in a basin pool used for watering outdoor green spaces.

面向足球场的立面有一个 3D 金属网围起来的通向公寓入口的人行通道。两栋楼在一层由一座窄桥连接，该桥也为学生们提供了一个开放的公共空间。

可持续性效率

该建筑的能效符合巴黎可持续发展的愿望。能源绩效目标和建造时间表汇聚在一个一年四季都发挥作用的简单、绝缘和通风良好的物体上。住房是交叉通风，并能让整栋公寓都能大量采光。外部走廊和玻璃楼梯也增强了通常循环中的自然采光，在创造舒适、明亮公共空间的同时提供能源。建筑外部用 20cm 厚的绝缘材料进行隔热。走廊地板和阳台用使用了热桥断路器以避免热桥。双流机械通风设备控制通风，常年为每个公寓提供最佳温度的干净空气，进来的空气也重用了从排气中的热量。屋顶上覆盖着 300 m² 的光伏板用于发电。雨水收集在一个盆池里用于浇灌户外的绿色空地。

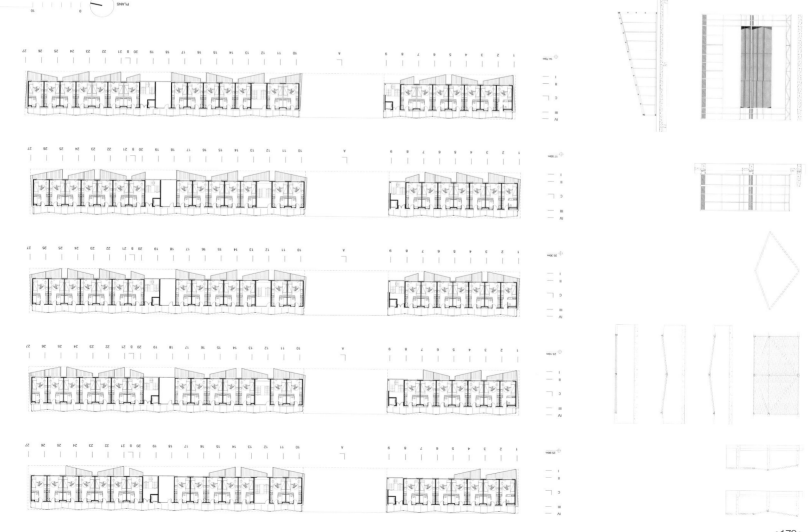

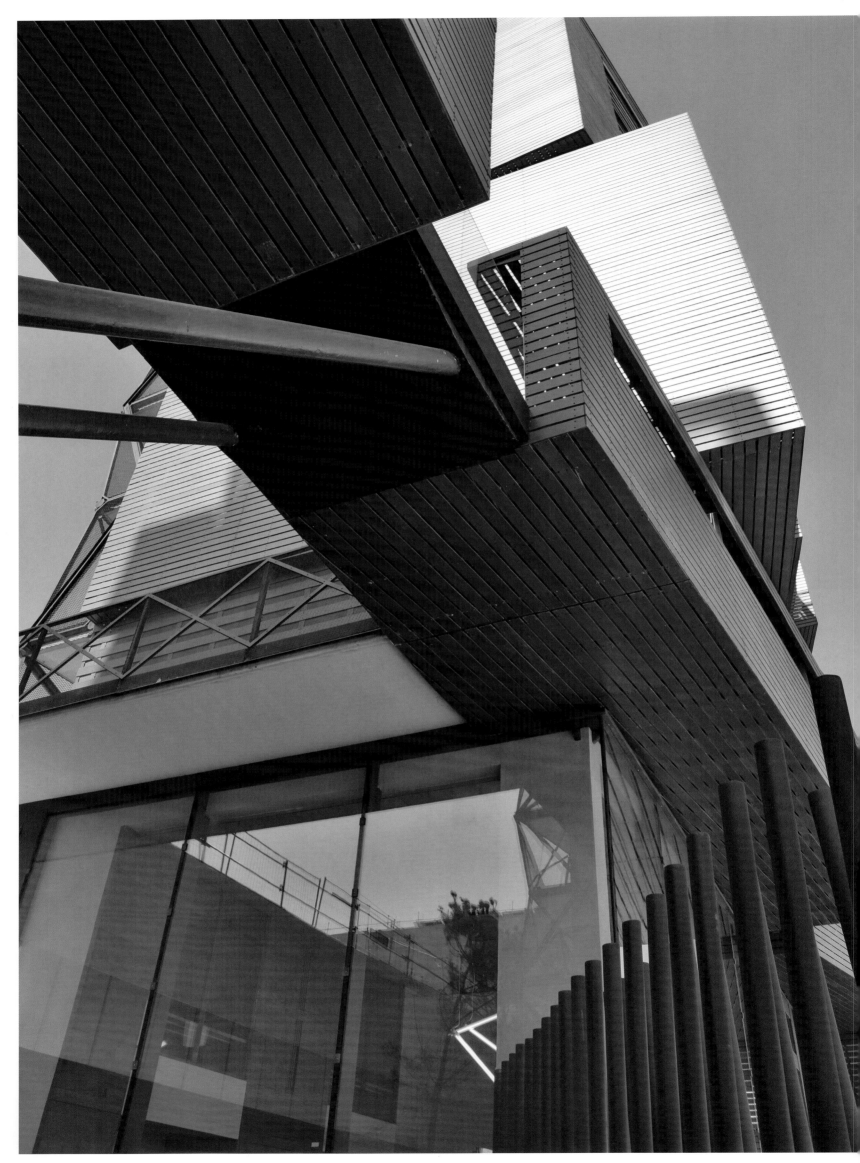

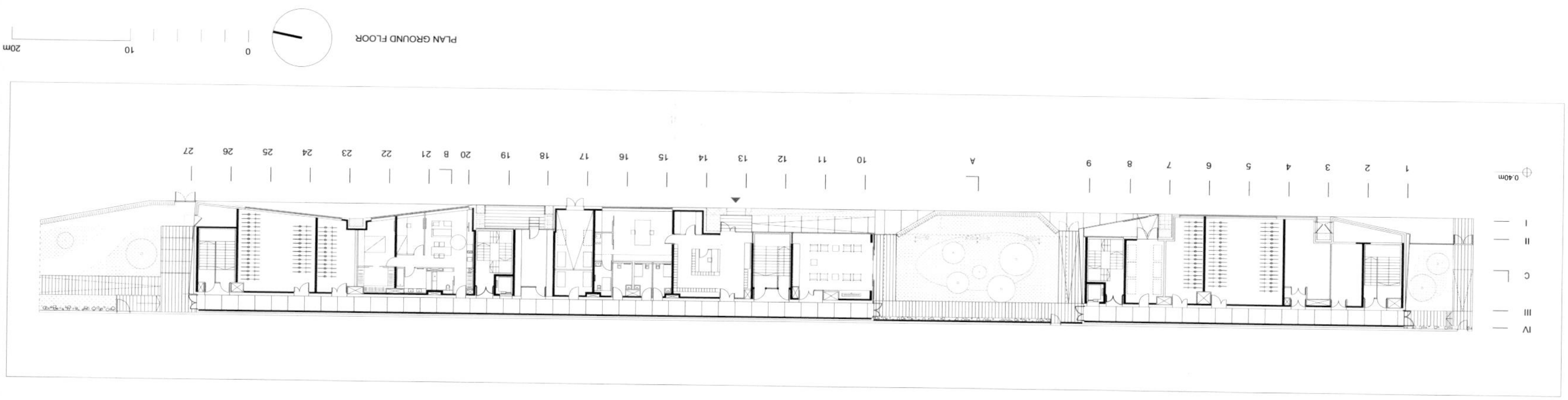

PLAN GROUND FLOOR

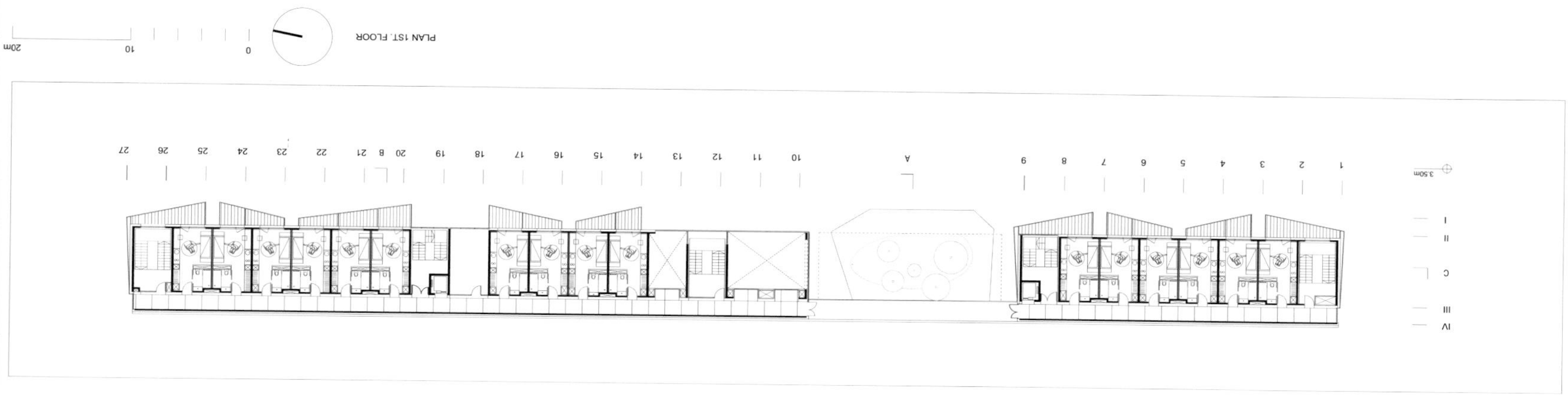

0   10   20m

27 26 25 24 23 22 21 20 B 19 18 17 16 15 14 13 12 11 10   A   9 8 7 6 5 4 3 2 1

I II C III IV   0.40m

PLAN 1ST FLOOR

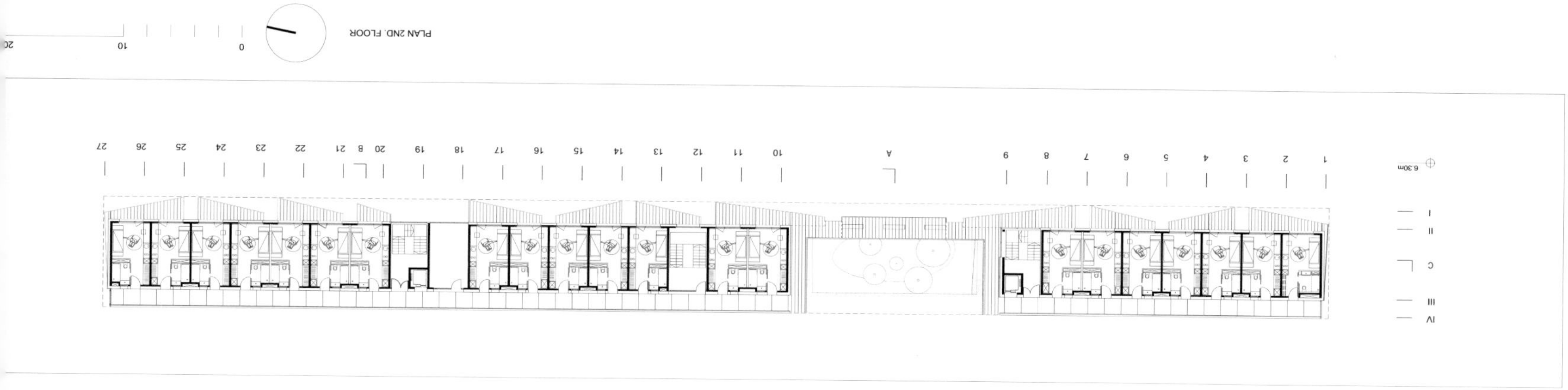

0   10   20m

27 26 25 24 23 22 21 20 B 19 18 17 16 15 14 13 12 11 10   A   9 8 7 6 5 4 3 2 1

I II C III IV   3.50m

PLAN 2ND. FLOOR

0   10   20

27 26 25 24 23 22 21 20 B 19 18 17 16 15 14 13 12 11 10   A   9 8 7 6 5 4 3 2 1

I II C III IV   6.30m

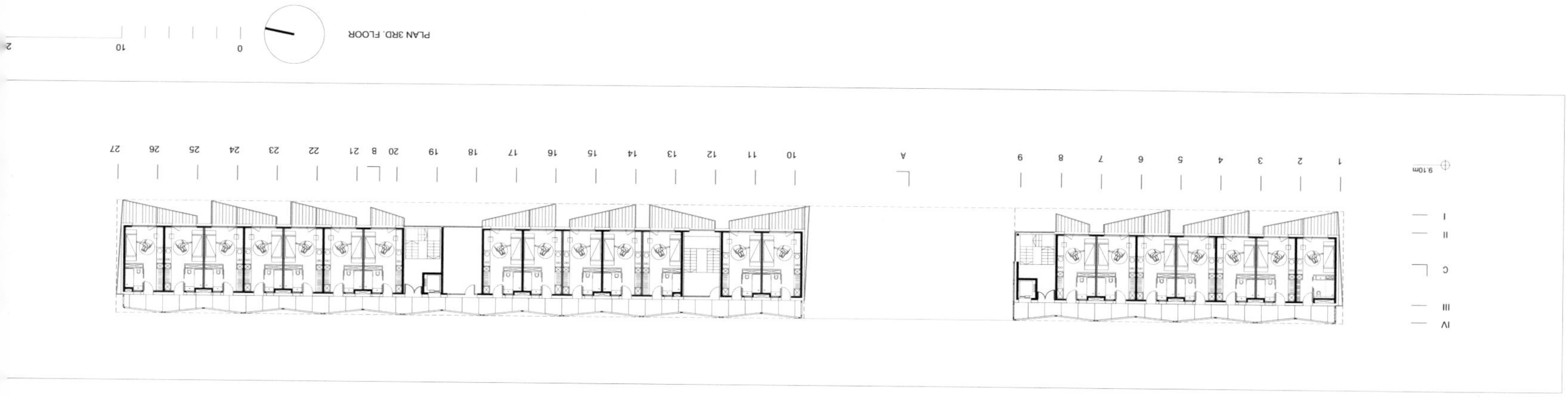

PLAN 3RD. FLOOR

0   10   2

27 26 25 24 23 22 21 20 B 19 18 17 16 15 14 13 12 11 10   A   9 8 7 6 5 4 3 2 1

I II C III IV   9.10m

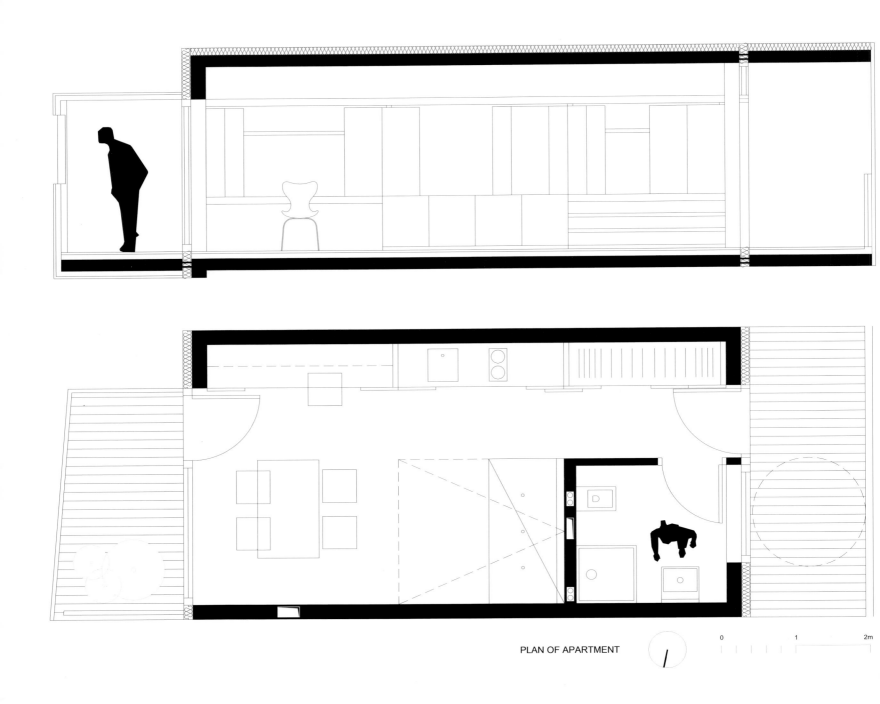

PLAN OF APARTMENT

0   1   2m

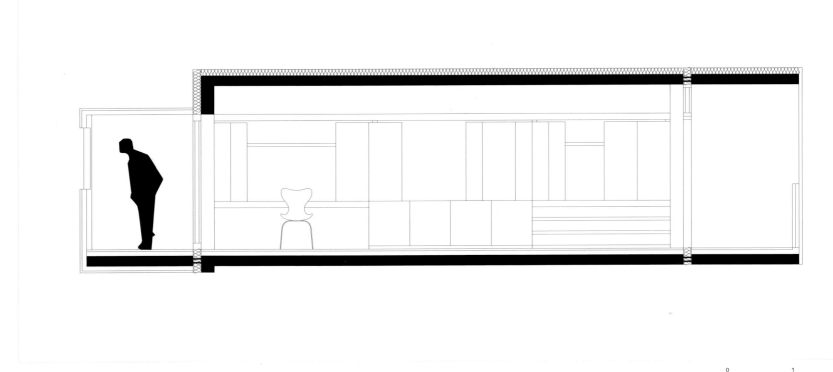

SECTION OF APARTMENT

0   1

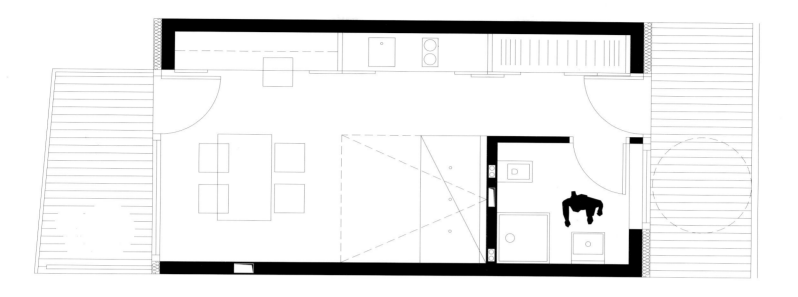

PLAN OF APARTMENT

0   1   2m

09
08
07
06
05
04
03
02
01
00

教师之家
Teachers Housing

This cluster of dwellings provides housing for primary school teachers in the small village of Gando. Six houses fan out in a wide arc from a shared arrival point, marking the southern limits of the school site. Three housing types, each based on a module as large as a traditional round hut, are combined in various ways to form a more complex whole.

The designs are simple and the range of materials minimized so that they can be

adopted and adapted by the villagers. The technology used in the houses is new in the region, climatically efficient and makes full use of local resources. No timber or steel was used. Each housing unit consists of three parallel walls made of stabilized earth brick. The 400mm thick walls stand on foundations of granite and concrete, which prevent moisture from rising.

　　此住宅群是为 Gando 村小学教师提供的住房。六栋房子从一个汇合点以扇形排开，标明了学校的南部界限。三种房屋类型，每种都在传统圆形小屋尺寸的模块上，以各种不同的方式进行组合以形成一个更复杂的整体。

　　该设计是简单的，材料范围也是最小化的，这样一来村民也可采用和改造。

住宅所用的技术在该地区是新型的，拥有气候效益并且充分利用了当地的资源。没有使用木材或钢材，每个住房单元由三面平行的墙壁组成，墙壁由稳固的土砖砌成。40cm 厚的墙建造在花岗岩和混凝土地基上，从而防止湿气渗入。

A tie beam at the top of the wall bears the loads from the barrel-vault. The roofs are constructed to two different heights. The intersection of the two forms a sickle-shaped opening, providing ventilation and daylight to the interior. Roof projections also protect the walls against erosion and moisture penetration, while specially formed channels at the top and ends of the walls allow for water runoff. In the render, bitumen replaced the traditional organic additives, giving a more durable finish. The housing project continues the principles of sustainable development and appropriate technologies established in the school building. Villagers assisted in the production of the building materials and the houses' construction.

墙顶的系梁承受了半圆形拱顶的荷载。屋顶被建造成两种不同的高度，两种高度的交叉处形成了镰刀形的开口，为室内提供通风和采光。屋顶突出物也可保护墙壁不受侵蚀和湿气渗透，墙顶两端专门形成的通道让水可以流出。在涂料方面，沥青代替了传统的有机添加剂，提供了一种更持久的表面。住房项目继续秉承这样的原则——可持续发展和在学校建筑中应用适合的技术。村民也参与了建材的生产和住房的建设。

Contributors / 设计师名录

Andrew Maynard Architects

Andrew Maynard has a Bachelor degree in Environmental Design and a Bachelor degree in Architecture with honours, both from the University of Tasmania. While still studying, Andrew won the Graphisoft international award for his design of Bulgakov's DEVIL'S BALLROOM. Following graduation he traveled extensively, pausing on rare occasions to work in offices such as Allom Lovell, Six Degrees and Richard Rogers. During this time Andrew entered and won both the Australia / New Zealand regional award and the overall Grand Prize in the Asia Pacific Design Awards for his DESIGN POD mobile workstation.

Upon returning to Melbourne in 2002 he established Andrew Maynard Architects with the intent of creating an office that gave equal value to both built projects and bold, polemical design studies. The resulting highly crafted built work and socio-political concepts have garnered global attention and recognition. Wallpaper* magazine, in its Architects Directory, an "annual guide to the world's most innovative practices",

offers this insight — "Andrew Maynard Architects are unafraid of mixing big conceptual designs with beautifully conceived small-scale works" while inhabitat says "In the diverse collection of projects we get a picture of an architect whose ideals run as deeply as his talent for good design." And, slightly less complimentary, the AGE states "His concepts include a man — eating robot, a bicycle made of plywood and "Poop House" — a structure made from human excrement and food. Images of the archetypal mad scientist spring to mind."

In 2004 Andrew was an awarded exhibitor at ADC Young Guns in New York. The Young Guns exhibition recognises international designers 30 and under. The following year Andrew's awarded prefabricated housing design received international attention for its push toward housing that is manufactured, as an industrial product, rather than built from sticks, plaster, brick and tin, offering a beautiful, fast affordable and sustainable housing solution.

artau Architecture

artau s.c.r.l. (Atelier de Recherche en Techniques spéciales, Architecture et Urbanisme (Research workshop for special techniques, architecture and urbanism) was founded in 1984 by five young architects: Luc Dutilleux, Norbert Nelles, Fabienne Hennequin, Serge Bonnevie and Marianne Misson.

Today, the architecture agency artau counts about sixteen persons including twelve architects, a draftsman, an assistant, an accountant and a communication manager.

The annual average turnover amounts to 1,5 million €, 40% of which come from the "Health care" sector, 40% from eco-friendly apartment complexes, the remaining 20% from the turnover of projects with different functions (apartment buildings, hotels, school buildings, restoration of patrimonial nature,etc).

artau s.c.r.l. claims above all an architecture that fits into context and presents an awakening effect.

The developed architecture intends to be a symbiosis with its context and the constraints that are linked to it.

The approach towards the client is in line with an altruistic will of awakening the architecture. It favours the active participation of the

client in each research phase so that the architecture perfectly meet the requirements and needs of every client.

Each of the projects handled by the artau s.c.r.l incorporates th technologies that are best suited for their function with a special focu on quality, user-friendliness, energy saving, respect of the environment.

For several years now this sustainable and environmental approach ha been at the heart of each project developed by artau architecture s.c.r.l.

In order to meet the constraints and objectives of their projects a best, artau architecture s.c.r.l. has built up over the years a networ of experts in different fields such as stability, special technique acoustics, urbanism, landscape architecture, lightning, environment conception, etc . Privileged partnerships with these experts enab artau to extend their sphere of competence beyond the genera mission that is generally carried out by an architecture agency.

artau architecture s.c.r.l. has always focussed on rigorous quality. Our ski improve with our audacious projects and thanks to the confidence of our client

Today artau architecture s.c.r.l. takes on new challenges by carrying out hug projects such as the new hospital in Liège and several housing complexe that respect the philosophy of eco-habitat and sustainable development.

BAK architects

BAK architects, is a studio established in 2000 and consisting of architects María Victoria Besonias, Guillermo de Almeida and Luciano Kruk.

The three owners have experience as teachers of architecture, so that each new assignment is an opportunity to implement the

link between theory and production as a way to work transcend the typecasting that often makes the exercise of the professio Thus since the request is presented begins a search for solutio with accompanying reflections framed in some now considere unavoidable questions.

baumraum

The architectural office baumraum, located in Bremen, Northern Germany, is specialized in the planning of experimental constructions on the ground, in trees and by or on water.

Since 2003 baumraum has planned and built inspiring dwellings for children and adults – from small playhuts for games and activities to exclusive permanent abodes. baumraum specializes in planning and realization of treehouses.

The constructions designed by baumraum are noted for their originality and inspiring creativity. Individual concepts are designed for private clients, hotels and catering businesses, forest environment and for special events. The safety and durability of the constructions are of primary importance.

The main concern thereby is to handle the trees and their

surroundings with the utmost care, ensuring their protection a preservation. The treehouses are not anchored by bolts or nails any other measures, which might injure the trees, but rather by text belts and adjustable steel cables, harmless to them.

The principal material used is domestic timber with its excelle properties like sustainability, water resistance and colour. But ev metal, textiles and plastics are potential materials for the archi tec baumraum combines the creative and constructive expertise architects with the long-standing experience of landscape designe tree experts, and established, reputable craftsmen.

The architects design suitable projects both in the natural and t urban environment. So far they have realized more than 40 projects Europe – Germany, Austria, Italy, Czech Republic and Hungary – k also in Brazil and the USA.

alderon-Folch-Sarsanedas Arquitectes

Ideron-Folch-Sarsanedas(CFS) Arquitectes is an architecture actice from Barcelona, founded in 2000 and leaded by Pilar Ilderon, Marc Folch and Pol Sarsanedas. Devoted to building, urban ojects, interior design, ephemeral staging and scenography, both in e international public and private field, CFS propose a "conscious chitecture" to improve people's quality of life focused on the lective interest through a socially and environmentally responsible

attitude. CFS has won numerous prizes in national and international competitions and deserved the Ajac Prize for young Catalan architects in 2004 and 2012. Pedestrian Bridge in the city centre of Manresa (2010), the Léonce Georges Cultural Centre in France (2012), the Low Energy MZ House in Barcelona (National Isover Energy Efficiency Awards 2012) and the Civic Centre, Library and Theatre in Begues (under construction) are some of their singular works.

-studio

chitecture is made for people by people and built within our natural vironment. We care about all 3. Our niche is in projects in which we n inventively combine technical and sustainable innovations with ong architectural qualities like remarkable spatial experience and cient functionality.

like to work for open minded clients and operate well with in the nstraints of tight budgets. Good constraints trigger more "out of the x thinking", more creativity, fun and inventiveness during the whole ocess. An open mind is important to be able to see and make use opportunities along the road. As a small studio we believe strongly

in an intelligent and interactive creative collaboration with all involved in the process of the design and its realization: clients, advisors, (sub) contractors and industries. Our office is situated in a multidisciplinary creative industrial environment www.b29studios.nl.

Our ongoing projects are: the construction of a small landscaped and green garden studio in Amsterdam, study for 30 greenhouses covered low-cost, low energy and adaptive housing (financed by the Dutch Innovation funds Agentschap NL and SEV foundation for experimental housing research in collaboration with the Technical University of Eindhoven) .

ébédo Francis Kéré

is the tradition in Burkina Faso, each member of the family is ponsible for the well-being of all members. Each individual is ispensable for the survival of the community. If one member leaves community in search of a better life, he tries to compensate his s by sending back financial support.

ncis Kéré wishes to fulfill his part of the social duty by offering a v perspective to the whole village. His projects reach way beyond ple financial support. His various stays in Europe have shown him t education and training are the basis of any social, professional d economic development. For this reason, our first goal was to ld a school in Gando, providing education to many children. Since n, several development projects have been set up to improve the ng conditions in the village.

ping people to help themselves constitutes the basis of all our jects. According to our philosophy, long-term considerations the priority for proper and sustainable development aid. In opinion, the key of sustainable development relies on the oulation's participation. The whole village community is involved he building process: the men manufacture the clay bricks, lay the

foundations, build up the walls and install the roofs. The women beat the clay floors and plaster the walls. Every morning for a whole year, the children bring a stone for the foundations on their way to school. The population's participation integrates the projects into the local communities and enables identification and motivation. For this reason, the results are valued, preserved and further developed.

Every innovation is first introduced into the school and afterwards into the society. Children accept change more easily than adults, and the next generations will naturally accept and use new facilities, for example latrines to prevent intestinal diseases.

We make sure that the constructions are as inexpensive and simple as possible. In order to ensure that the population is able to use similar techniques, the buildings are made of clay which is found everywhere in the region. The extreme climatic conditions of Burkina Faso require a particular and adapted architecture. Our construction method, using natural ventilation to cool down the room climate, is appropriate for the temperature. The overhanging roof, the clay walls and the natural cooling system prevent the building from overheating. The whole country is currently speaking of the benefits of this way of constructing.

anz Architekten

the beginning of 2009 Robert Diem and his colleague Erwin ttner founded the architekture-company franz zt gmbh.

a young office without a fixed job they were forced to attend olic competitions, which led to a first job fortunately, a upper ondary school with focus on natural sciences and mathematics ealgymnasium") and at the same time the existing secondary ool ("Hauptschule") in deutsch-wagram needed to be expanded

with € 10,5 mio. building costs.

In the last few years Franz did 30 competitions, received 15 prices, with four 1st prices. In June 2010 they won two more competitions about 50 participants: upper secondary school ("Gymnasium") in Gainfarn € 14 mio. building costs and provinical youth hostel ("landesjugendheim") in Hollabrunn € 7,5 mio. building costs.

GRAFT

The English word "graft" provokes a variety of meanings and multiple readings. In the terminology of botany, grafting is described as the addition of one shoot onto a genetically different host. The positive properties of two genetically different cultures are combined in the new biological hybrid.

GRAFT, established in 1998 in Los Angeles, California by Lars Krückeberg, Wolfram Putz, Thomas Willemeit and Gregor Hoheisel, is as a "Label" for architecture, urban planning, design, music and the "pursuit of happiness". With further offices in Berlin and Beijing,

GRAFT has been commissioned to design and manage a wide range of projects in multiple disciplines and locations. With the core of the firm's enterprises gravitating around the field of architecture and the built environment, GRAFT has always maintained an interest in crossing the boundaries between disciplines and "grafting" the creative potentials and methodologies of different realities. This is reflected in the firm's expansion into the fields of exhibition design and product design, art installations, academic projects and "events" as well as in the variety of project locations.

harunatsu-arch

Harunatsu-arch is an architectural practice based in Japan founded by Shoko Murakaji and Naoto Murakaji in 2011 after they left Tezuka Architects. They have office in Kanazawa, which is a historical town located in the central part of mainland Japan. Their office itself is the former Samurai residence in the most historical quarter in Kanazawa.

The work by harunatsu-arch can be featured in their continuous attempt to change the modernism architecture through dialog with

local climate and context. Although their work does not appear to be vanacular at a glance, it consists of various ideas, which makes the architecture sustainable within the local environment.

Harunatsu-arch's approach is based on an informed use of essence of the traditional houses with a strong awareness of local context then creating architecture, which is appropriate for today's age.

James & Mau

James & Mau is an international architecture firm based in Madrid that brings together the expertise in several areas such as architecture, construction and real estate development in order to offer a full service support for the development of the projects both for individuals and enterprises.

The company is widely experienced in design, work on-site and building permits, with a strong focus on modular industrialized architecture and construction, applying bioclimatic and sustainable concepts.

One of the added values of James & Mau is its international profile with a strong knowledge of Latin American market, especially of Chile and Colombia where it has its offices. The firm aims to position itself as a platform for European investors overseas.

James & Mau was founded in 2007 by Jaime Gaztelu and Mauricio Galeano, and in turn, participates in Infiniski, company dedicated to modular and sustainable construction very much concerned with high quality design, fast execution and budget accuracy.

JM Architecture

JM Architecture, founded in 2005 by Jacopo Mascheroni and based in Milan, provides a range of architectural and design services to clients in Italy and abroad. The firm creates spaces where refined, pure, and timeless architectural lines meet with the most advanced technology to provide a graceful combination of exceptional aesthetic elegance, utility, and comfort. Each project is approached as a unique opportunity and is a tailor-made solution incorporating the firm's meticulous attention to details, finishes, and materials selection. In both new-construction and renovations, the appropriate integration

of the architecture into the site and the surrounding landscape a priority, and careful implementation of home-automation, audio video systems and energy efficient solutions are hallmarks of each project. The works have earned national and international attention and have been published in print on four continents as well as widely published online. The dynamic output of the studio is driven by the collaboration of the firm's talented professionals who represent many nationalities and backgrounds and continuously strive to imbue the work with the ideals of simplicity, coherence, clarity and harmony.

KUCZIA / Peter Kuczia

Born in Poland. Studied architecture at Silesian University of Technology, Poland. 2008 dissertation on solar energy use in architecture. Since 1999 he has worked for agn Group in Ibbenbüren, Germany and as freelancing

architect in Poland. He has been awarded several international prizes. His work – which covers a wide range of building types – has been exhibited and published around the world.

OFIS Architects

OFIS Architects is an architectural practice established in 1996 by Rok Oman (1970) and Špela Videnik (1971), both graduates from the Ljubljana School of Architecture (graduated in October 1998) and London's Architectural Association (MA in January 2000). By 1998 they had already won several prominent competitions, such as Football Stadium Maribor and City Museum Ljubljana. Many of their projects have been nominated for a Mies van der Rohe Award, a Silver IOC/IAKS medal for the football stadium in 2009, the European Grand Prix for Innovation Award in 2006, an honorable mention at the Miami Bienal in 2005 for their Villa "Under" Extension, a high commendation for the City Museum renovation and extension by the UK Architectural Review's AR+D Awards, and the prestigious "Young Architect of the Year" Award in London, UK, in 2000, to just mention a few of their achievements.

The international team is based in Ljubljana, Slovenia and Paris, France and has partner firm agreements in London and Moscow. The beginnings of OFIS' activities date back to the nineties, a particularly exciting yet difficult period for the former Yugoslavian republics that were undergoing intense self-re-evaluation and reinvention from scratch, economically and culturally. In terms of architecture this meant that most of the larger architectural offices had to be scaled down or went bankrupt, creating an empty space for younger groups or individuals to participate in architectural competitions. Back then, OFIS managed to impress the jurors with original thinking and clear concepts. Over the past ten years they have been dealing with various national and international clients from the private sector, the commercial sector, and state institutions- each with their own set of agendas, budgets and problems. In this period, they have developed the tactics in their work which make them stand out from the rest.

PES-Architects Ltd.

PES-Architects is one of the leading and most international architectural design firms in Finland. Professor, Architect Pekka Salminen founded the company in 1968, giving the office over 40 years of continuous success.

The main projects of PES-Architect are such complex public buildings as theatres, concert halls, airport and railway terminals, but also schools and sports facilities, retail developments and office buildings.

The partners of PES-Architects are Pekka Salminen and Tuomas Kivennoinen as Main Designers and Jarkko Salminen as GEO.

Besides architectural design, the line of activities includes interior design, urban planning and project management.

PES-Architects has operated in China since 2003 and in 2011 the Chinese company PES-Architects Consulting (Shanghai) Co., Ltd., P.R. China was established.

The main realized projects in Europe are Helsinki Airport in Finland and St Mary Concert Hall in Germany.

PES-Architects has participated in 60 architectural competitions in China. The Wuxi Grand Theatre was inaugurated in April 2012. The construction of the Chengdu Icon Yuan Duan super high-rise tower started at the beginning of 2011.

About ARTPOWER

PLANNING OF PUBLISHING

Independent plan, solicit contribution, printing, sales of books covering architecture, interior, graphic, landscape and property development.

BOOK DISTRIBUTION

Publishing and acting agency for various art design books. We support in-city call order, door to door service, mail and online order etc.

COPYRIGHT COOPERATION

To further expand international cooperation, enrich publication varieties and meet readers' multi-level needs, we stick to seeking and pioneering spirit all the way and positively seek copyright trade cooperation with excellent publishing organizations both at home and abroad.

PORTFOLIO

We can edit and publish magazine/portfolio for enterprises or design studios according to their needs.

BOOKS OF PROPERTY DEVELOPMENT AND OPERATION

We organize the publication of books about property development, providing models of property project planning and operation management for real estate developer, real estate consulting company, etc.

INTRODUCTION OF ACS MAGAZINE

ACS is a professional bimonthly magazine specializing in high-end space design. It is color printing, with 168 pages and the size of 245*325mm. There are six issues which are released in the even months every year. Featured in both Chinese and English, ACS is distributed nationwide and overseas. As the most cutting-edge counseling magazine, ACS provides readers with the latest works of the very best architects and interior designers and leads the new fashion in space design. "Present the best whole-heartedly, with books as media" is always our slogan. ACS will be dedicated to building the bridge between art and design and creating the platform for within-industry communication.

Artpower International Publishing Co., Ltd.

Add: G009, Floor 7th, Yimao Centre, Meiyuan Road, Luohu District, Shenzhen, China
Contact: Ms. Wang
Tel: +86 755 8291 3355
Web: www.artpower.com.cn
E-mail: rainly@artpower.com.cn

QR (Quick Response) Code of ACS Official Wechat Account

Acknowledgements

We would like to thank all the designers and companies who made significant contributions to the compilation of this book. Without them, this project would not have been possible. We would also like to thank many others whose names did not appear on the credits, but made specific input and support for the project from beginning to end.

Future Editions

If you would like to contribute to the next edition of Artpower, please email us your details to: artpower@artpower.com.cn